Literatures, Cultures, and the Environment

Series Editors
Usrula K. Heise
University of California,
Los Angeles
Los Angeles, USA

Gisela Heffes
Johns Hopkins University
Baltimore, USA

Literatures, Cultures, and the Environment focuses on new research in the Environmental Humanities, particularly work with a rhetorical or literary dimension. Books in this series explore how ideas of nature and environmental concerns are expressed in different cultural contexts and at different historical moments. They investigate how cultural assumptions and practices, as well as social structures and institutions, shape conceptions of nature, the natural, species boundaries, uses of plants, animals and natural resources, the human body in its environmental dimensions, environmental health and illness, and relations between nature and technology. In turn, the series makes visible how concepts of nature and forms of environmentalist thought and representation arise from the confluence of a community's ecological and social conditions with its cultural assumptions, perceptions, and institutions.

Rod Giblett

Wetland Cultures

Ancient, Traditional, Contemporary

palgrave
macmillan

Rod Giblett
Writing and Literature Program
Deakin University
Melbourne, VIC, Australia

ISSN 2946-3157 ISSN 2946-3165 (electronic)
Literatures, Cultures, and the Environment
ISBN 978-3-031-57364-4 ISBN 978-3-031-57365-1 (eBook)
https://doi.org/10.1007/978-3-031-57365-1

Cover illustration: Southern Lightscapes-Australia \ gettyimages

This Palgrave Macmillan imprint is published by the registered company Springer Nature Switzerland AG.
The registered company address is: Gewerbestrasse 11, 6330 Cham, Switzerland

Paper in this product is recyclable.

Mary Oliver, 'Crossing the Swamp' 1980
Here is the endless
wet thick
cosmos, the center
of everything—the nugget
of dense sap, branching
vines, the dark burred
faintly belching
bogs. Here
is swamp, here
is struggle,
closure—
pathless, seamless,
peerless mud. My bones
knock together at the pale
joints, trying
for foothold, fingerhold
mindhold over
such slick crossings, deep
hipholes, hummocks
that sink silently
into the black, slack
earthsou I feel
not wet so much as
painted and glittered

with the fat grassy
mires, the rich
and succulent marrows
of earth—a poor
dry stick given
one more chance by the whims
of swamp water—a bough
that still, after all these years,
could take root,
sprout, branch out, bud—
make of its life a breathing
palace of leaves.

First published in The Atlantic, *May 1980.*
Republished in American Primitive *(Little, Brown, 1983).*

Dedicated to wetlands—marginalized communities, oppressed minorities

Preface and Acknowledgments

The present volume had its beginnings in a number of serendipitous events. I am grateful to John Charles Ryan for drawing my attention to the term 'paludal' when he described swamp lovers and wetland conservationists as 'paludal heroes.' His and other usages of paludal are discussed in the introduction to the present volume. John also provided an electronic copy of Federico Borca's chapter on towns and marshes (*paludes*) in ancient Rome for which I am grateful. I had already come across Borca's article on ancient Roman wetland culture. John also pointed me to Mary Oliver's poem 'Crossing the Swamp' reproduced as the epigraph to the present volume. It is gratefully reproduced with the permission of the Licensor through PLSclear.

Ryan's and Borca's work stimulated my curiosity about the paludal in the classical Latin canon. John encouraged me to explore it more for which I am grateful too. I read for the first time Cato, Tacitus, Varro and Virgil in the bi-lingual Loeb Classical Library editions with translations in parallel text so I could pick up the paludal and its declensions in the original Latin. I discussed Pliny and Lucretius briefly in *Postmodern Wetlands* published in 1996. I also discussed Plato, Aristotle, Hippocrates and Homer briefly and intermittently in *Postmodern Wetlands*. In Chap. 4 of the present volume, I draw together and develop the discussion of these ancient Greeks. For the present volume, I read more of Pliny's work in the Loeb edition with its translation. I also read a recent translation of Lucretius. In Chap. 5 of the present volume, I elaborate on them and all the ancient Roman writers mentioned. I am not a classicist so I hope any missteps on my part are forgiven.

The present volume also had its beginnings in recent research on ongoing topics of interest. I recently read Eric Ash's histories of the draining of the English Fens and discuss them in Chap. 4. An earlier version of Chap. 5 on the culture of the Fens was first published in *Wetlands and Western Cultures: Denigration to Conservation* published in 2021. I am grateful to the publisher and copyright holder, Rowman & Littlefield, for permission to reproduce already published material with all rights reserved.

I have previously discussed American wetland cultures, especially of the Great Dismal Swamp between Virginia and North Carolina, the major North American wetland cities, Henry David Thoreau as the patron saint of swamps and Aldo Leopold as the patron saint of marshes. My attention in the present volume turns across the northern border in North America to Canada and its marshy provinces of New Brunswick and Nova Scotia. I have also previously discussed Aboriginal and non-Aboriginal wetland cultures in Australia in the states of Victoria, Western Australia and New South Wales. In Australia, my attention turns mainly to the state of Queensland and briefly to South Australia and elsewhere in Victoria. The present volume takes a transnational and transhistorical point of view of wetland cultures.

The wetlands of the Murray-Darling Basin spanning four states, including Queensland, South Australia and New South Wales, have recently been Emily O'Gorman's focus in *Wetlands in a Dry Land: More-Than-Human Histories of Australia's Murray-Darling Basin* published in 2021. When I first read it I was surprised that the chapter on pelicans and the Coorong wetlands (where the Murray River meets the sea in South Australia) does not discuss Colin Thiele's classic *Storm Boy,* which is about both (among other things) and which has been made into a film twice. Such is the wont of environmental historians in their siloed discipline not to discuss film and literature that inform wetland cultures in the past and present and for the future. By contrast, the transdisciplinary environmental humanities in general, and wetland cultural studies in particular, embrace film, literature, history and philosophy, as well as employ feminist, Marxist, psychoanalytic and post-structuralist approaches. The present volume employs these approaches and embraces these disciplines and media. Writing the present volume gave me the opportunity (for which I am grateful) to discuss Thiele's *Storm Boy* first published in 1963 and his later *Swan Song* published in 2002 about pelicans and the Coorong wetlands too (as well as a black swan).

When I Googled *palus,* 'palustrine' came up in the recent book *Wetlands of Queensland* published by the Queensland Museum in 2022. When I visited the Brambuk Cultural Centre in Halls Gap in western Victoria in March 2023, I came across James Morrill's account of living among Aboriginals in the swamps of north Queensland for 17 years in the mid-nineteenth century. I remembered that Judith Wright, who lived much of her life in Queensland and is the patron saint of Australian conservation, wrote a word for wetlands. I am grateful to John Charles Ryan for drawing my attention to her poem 'Swamp Plant.' I also remembered Alexis Wright's two novels, *Carpentaria* published in 2006 and *The Swan Book* published in 2013, set in the main in Queensland (and in lagoons and swamps in part, as it turned out when I read them). I am grateful to Emily Potter for recommending *The Swan Book* to me a long time ago as I had also written about black swans at the same time in *Black Swan Lake* also published in 2013. I am grateful to Judy Gunning of the Queensland Art Gallery Museum of Modern Art for drawing my attention to Mavis Ngallametta's two paintings of Queensland lagoons or swamps held in the gallery and to the Winton wetlands sculpture walk in Victoria. I am also grateful to the gallery for permission to reproduce Conrad Martens' painting *Coochin Run* and to the Mitchell Library of the State Library of New South Wales for Martens' painting *Coochin 27 Nov. 1851.*

While researching the present volume, I used 'Trove,' the online data base of Australian libraries, entered 'Queensland swamps' in its search engine and found the newspaper article by 'Mopoke,' the chapter by J. W. Winter and Curzon-Siggers' history of Cairns. I am grateful that the current Australian government granted special financial support to sustain 'Trove' when it was in danger of closing. This would have been an unmitigated disaster for researching Australian history and culture and would have meant the continued overlooking or neglecting and inaccessibility of important aspects of both, such as wetlands. I draw all these topics together about Queensland wetlands for discussion in the second chapter of the present volume

A similar approach to wetland cultures to that undertaken in the present and previous volumes remains to be undertaken in other states and the territories of Australia with further research and writing, such as the Northern Territory. While researching the present volume I came across references to a wealth of tantalising and mouth-watering materials held in the Queensland Art Gallery relating to Aboriginal wetland cultures in the Northern Territory that would reward research and discussion. These include:

Port Keats (wadeye) Community / Murrinh-patha people / *Rainbow serpent story* c.1962 / Watercolour on bark glued to table top / 54.5 × 15.5 cm (oval); table top: 62.4 × 23.9 × 2.9 cm / Gift of G.W. Spence 1971 / Accession No: 1:1173;
Abie Jangala / Warlpiri people c.1919 - 2002 / *Water Dreaming* 1994 / Synthetic polymer paint on canvas / 174.1 × 144.6 cm / Purchased 1995. / Accession No: 1995.075; and
Yirrkala Community / Yolngu people, Arnhem Land / *Swamp* 1948 / Natural pigments on paper / 58.5 × 45.7 cm / Gift of the 1948 American-Australian Scientific Expedition to Arnhem Land 1956 / Accession No: 1991.370.

Images, including thumbnails, of these artworks are not currently available on the Queensland Art Gallery of Modern Art website.

South Australian wetlands require similar research and discussion too. They are briefly touched on in the first, introductory chapter of the present volume in relation to John Olsen's famous paintings of Lake Eyre-Kati Thanda and Colin Thiele's classic children's books set in the Coorong Wetlands (as already mentioned). Elsewhere I have previously discussed James Ashton's and Henry James Johnstone's paintings of South Australian wetlands, both of which are housed in the Art Gallery of South Australia. No doubt there is more to find and more to do with Australian wetland cultures in South Australia and the Northern Territory, as well as elsewhere in Australia, such as in Tasmania and the Australian Capital Territory.

The Marsh Arabs of Iraq have been lurking in the back of my mind for some years so I decided it was time to follow up on this, beginning with reading Wilfred Thesiger's classic *The Marsh Arabs* readily available in a recent Penguin Classics edition with an updated introduction. I had briefly referred to James Scott's discussion of Jennifer Pournelle's work on the Marsh Arabs in a previous book so I tracked down her primary text. I am grateful to Robert Thorson, the patron saint of Thoreauvian students, for drawing my attention to Scott's book. I also came across recent and very recent work on the marsh Arabs in the garden of Eden. The ducks were lining up. I am grateful to Steve Lonergan for providing a pre-publication PDF of his and Jassim Al-Asadi's book, *The Ghosts of Iraq's Marshes* published in April 2024 that brings the story of the Marsh Arabs and their marshes right up to date with an insider's view. I do not read Arabic so I have relied on translations into English and other English language studies of the marsh Arabs and their marshes.

I am grateful to the government and taxpayers of Canada for awarding me an 'Understanding Canada' grant that enabled me to travel to and across Canada and to visit many wetlands, including in the marshy provinces of New Brunswick and Nova Scotia, aka Acadia. I am grateful to Ed and Mary Ann Sulis for their hospitality in Kentville, Nova Scotia and Ed for his guided tour of the Grand Pré dykes and marshlands. I am also grateful to Anusha Beechowa for translating Andrew Gann's chapter on Antonine Maillet's novels set in Acadia from French into English and David Elder for arranging for Anusha to do the translation. The marais Acadians of Canada are discussed in Chap. 8 of the present volume.

Ancient, traditional and contemporary wetland cultures are also present in other countries with their wetlands beside Australia, England, Ireland and Canada discussed in the present volume. These include the major wetlands of the Okavango Delta in Botswana and the Pantanal in Brazil. Both wetland cultures are outside my expertise and access to resources to research, as well as being beyond the scope and word-limit of the present volume. Wetland cultural studies similar to the present volume may have already been undertaken of these and other wetland cultures, or may remain to be undertaken.

I am not an Aboriginal person so I do not speak, nor claim to speak, with any cultural authority about Aboriginal cultures. I rely on published sources. I am just a white-fella trying to learn from, and engage in dialogue with, Aboriginal peoples via the written word and to share that learning with non-Aboriginal peoples. I am not Arabic, Acadian, British, Irish, Greek or Italian. I do not speak, nor do I claim to speak, with any cultural authority about their cultures. I rely on translations into English and studies in English. I am a wetland lover (paludiphile) and student of wetland cultures who lived by a wetland for 28 years and wrote about it in several books. I also wrote about other wetlands and their cultures in other books. I am a promoter of pro-paludal wetland cultures and paludiphilia—love of wetlands.

Melbourne, VIC, Australia
December 2023

Rod Giblett

Praise for *Wetland Cultures*

"With more than 50% of the human body being water, it should be expected that a visceral connection with wetlands runs deep through our diverse cultural histories. Yet much of this has been smothered by marginalising narratives. With politically sharp anthropological scholarship and the author's personal passion, *Wetland Cultures* recalibrates our understanding of these watery environments, and at a time of ecological crisis, it provides essential reading for enthusiasts and policy-makers alike."

—Dave Pritchard, Coordinator, *Ramsar Culture Network, UK*

"This beautifully composed and curated work is as supple and serpentine as the waterscapes it explores. The reader is taken on a wonderous immersion into the cultural practices and affiliations of a myriad of continuing planetary paludal encounters between humans and wetlands. Encompassing bodily and spiritual entanglements with these complex and dynamic ecosystems, Giblett affords us a glimpse into other worlds running parallel with our modern lives, to question our relationship with the ancients, with landscapes, with each other. A delicate and joyous watery, marshy sojourn into ways of knowing, seeing and being that questions the very essence of our collective lives together as co-connected biota."

—Mary Gearey, *Senior Lecturer in Human Geography, School of Applied Sciences, University of Brighton, UK*

"Rod Giblett presents a fascinating, original and thought-provoking account of how wetlands have shaped people and culture throughout history. From an exploration of wetland representations in art and literature, to their environmental, spiritual and agricultural values that often reflect the colonial gaze, the book offers critical insight into this rich cultural heritage—the implications of which continue to be overlooked in the mainstream global wetlands discourse."

—Alan Dixon, *Professor of Sustainable Development, University of Worcester, UK*

Contents

About the Author

Rod Giblett is the author of 30 books of fiction and faction ('non-fiction'), including recently *Middlemarsh: The Hopkins River, Kindred Wetlands and Remarkable People* and *The Way of Taoism: For Bodily, Spiritual and Environmental Health* (both Transnational Press, 2023). He lived by a wetland in Western Australia for 28 years, was active in its conservation for 25 years and wrote about it in many articles and several books. He now lives in Melbourne, Australia and wrote about its wetlands too. He is a graduate of the Universities of Western Australia and Sydney and Murdoch University. He taught and researched at three universities in Western Australia for 25 years, including travelling to Canada on study leave, visiting many of its wetlands and researching and writing about them. He is the founder of wetland cultural studies, psychoanalytic ecology, conservation counter-theology and Thoreau and Benjamin studies. He is Honorary Associate Professor of Environmental Humanities in the Writing and Literature Program at Deakin University, Australia.

List of Figures

CHAPTER 1

Introduction to Wetland Cultures, Past and Present

Traditional cultures have a long and vital association with wetlands going back to ancient times and coming forward to contemporary time. They use paludicultural and aquacultural techniques to provide sustainable sustenance (rather than merely 'subsistence'). Paludiculture comes from the Latin word '*palus*,' meaning 'mire' or 'marsh.' It is in current usage to refer to the cultivation of the peatlands of England and northern Europe in the past and present and for the future (Milner and Stuart 2022; Wichtmann et al. 2016). The history of the use of peatlands in northern Europe has been researched and traced recently by Ruuskanen (2016). I use the term paludiculture more broadly to refer to the cultivation of plants for animal and human use and consumption in all types of wetlands, including bogs, lagoons, marshes and swamps. I define paludiculture as the cultivation of local, native (endemic) plants in wetlands for animal and human use and consumption, including eating and heating. Paludiculture forms the basis for the broader expression of wetland culture in artefacts (canoes, spears, sticks, nets, baskets, houses, household vessels, etc.) and the arts, including story-telling, dance, writing, painting, textiles, jewellery and basket-weaving.

'*Palus*' is also incorporated into the common word 'palusplain' where it is used to refer to a seasonally waterlogged, flat wetland. It is in current usage to describe much of the Swan Coastal Plain of south-western

R. Giblett, *Wetland Cultures*, Literatures, Cultures, and the Environment, https://doi.org/10.1007/978-3-031-57365-1_1

Australia encompassing Perth, 'a city of wetlands,' including palusplains.[1] 'Paludal' has been used recently to describe swamp lovers and wetland conservationists as 'paludal heroes,' as Ryan and Chen (2020) call them. 'Palustrine' has been used recently by the Queensland Museum (2022, 7) to describe 'primarily vegetated non-channel environments with more than 30% emergent vegetation.' Examples of palustrine wetlands include 'grass, herb and sedge swamp; wet heath swamps (wallum); *Melaleuca* spp. and *Eucalyptus* spp. tree swamps' and other swamps (Queensland Museum 2022, 7). Aboriginal people have cultivated these and other wetlands in Queensland for thousands of years. More than mere 'swamp managers' and wetland conservationists (though they were this too) they were wetland cultivators. Paludiculture of palustrine and other types of wetlands in Queensland, such as lagoons and mangroves, is discussed in the second chapter of the present volume.

1.1 Traditional Wetland Cultures

The peoples of traditional wetland cultures regard wetlands as sacred places imbued with spiritual and ceremonial significance (Verschuuren 2018). They also provide physical sustenance and are sources of materials. These values are expressed in their artworks, artefacts, dance, song and story (Gearey and Giblett forthcoming). Their artefacts, such as baskets, canoes and spears, were, and still are, created from wetlands and used in wetlands to hunt and gather the food they cultivate there. For example, Aboriginal peoples practised 'fire-sticking farming' and used 'digging sticks' in the wetlands on the fertile plains of mid-western Victoria in Australia to cultivate food for themselves, such as the yam daisy in paludiculture, and to cultivate pasture of native grasses, such as kangaroo grass, for grazing mobs of native kangaroos and wallabies (what I have called pasturalism, as distinct from pastoralism, the herding of introduced sheep and cattle on drylands; see Giblett 2023, 25, 67). This material circle and seasonal cycle of sustenance sustained their livelihoods and lives for tens of thousands of years.

Aboriginal people also engineered water and stone in mid-western Victoria. Eel and fish traps and stone houses located at Lake Condah and

[1] For further discussion of Perth, the capital city of Western Australia, set beside the (black) Swan River and among wetlands, including palusplains, see Giblett (1996, Chap. 3; 2013, Chap. 15).

Condah Swamp in the country of the Gunditjmara people date back over 4000 years. These structures are older than Stonehenge (Gunditjmara People with Wettenhall 2010). The traps are 'the world's oldest and largest aquaculture system' (Langton 2021, 15, 65–66). The Budj Bim site in the wetlands of Lake Condah (Gunditjmara People 2010), was gazetted as National Historic Landscape of Australia in 2004 (Cahir et al. 2018, 283) and is now a World Heritage Site (AAP 2019; Langton 2021, 13–17, 409). More than intensive hunters, gatherers, fishers and foragers, the Gunditjmara people of mid-western Victoria around Lake Condah were designers and builders of eel and fish traps (or 'engineers of aquaculture'; Gunditjmara People 2010, 16–22) and cultivators of wetland plants (Gunditjmara People 2010, 7, 13–16, 67), or practitioners of paludiculture. These people practised architecture with the construction of stone houses, agriculture with the cultivation of native grasses and cereals, aquaculture with the building of stone traps and weaving of baskets to catch eels and fish, aquaorniculture (from the Latin *aqua* for water and the Greek *orni* for bird) with the weaving of nets to snare water birds and paludiculture with the cultivation of edible wetland plants. More than mere 'swamp managers' and swamp farmers for the engineering of water in aquaculture and in irrigation for the cultivation and harvesting of edible plants and animals in paludiculture, they were also paludiphiles, lovers of wetlands.

These practices were the basis and indicators of Aboriginal civilization as they involved 'making improvements' to use John Stuart Mill's definition of civilization (cited by Giblett 2023, 25). Aboriginal people also demonstrated ownership of the land as they 'mixed their labor with nature' to use John Locke's definition (cited by Giblett 2023, 26). Applying these definitions of civilization and land ownership to Aboriginal peoples demolishes the doctrine of *terra nullius*, meaning 'nobody's land,' and so land legally uninhabited. It was applied most famously to Australia by Captain James Cook who could then claim and proclaim British royal sovereignty with impunity. It denies the work of Aboriginal hunter-gatherers, nomads, architects of stone houses, engineers of aquaculture, pyrotechnicians of the bush, cultivators of grasslands and wetlands, graziers of pasturalism and so farmers of agriculture, paludiculture and aquaorniculture. In all of these practices, they mixed their labour with nature, demonstrated their ownership and inhabitation of the land, expressed their civilization in it and proclaimed their sovereignty over it. Australia 'always was, always will be' Aboriginal country.

Many aspects of traditional wetlands cultures are contemporary in the sense that they endure in wetland Aboriginals', Marsh Arabs', Fen Britons', bog Irish and marais Acadians' appreciation of wetlands as sacred places of spiritual significance and as vital ecosystems, food sources and habitats for human and more-than-human life on earth. All of these people practise paludiculture, the cultivation of wetlands. Their cultures are pro-paludal, for wetlands. Traditional attitudes to wetlands also endure in writing and artworks created over the past two centuries in Aboriginal wetland peoples' portrayal of swamps and the depiction of the British Fens. A traditional wetland culture also persists among the Marsh Arabs of southern Iraq, despite decades of destruction and genocide. Traditional Irish attitudes to bogs as the source of peat and as the landscape of national identity endure in twentieth-century writing about them. Traditional French attitudes to marshes as good grazing land with managed water flows persist and are visible in Acadia (French Canada) and in writing about them.

1.2 Ancient Greek and Roman Cultures

By contrast, ancient Greek and Roman cultures denigrated (literally and figuratively 'blackened') wetlands as places of disease, terror, horror, the hellish and the monstrous. These attitudes are anti-paludal and placist in that they ascribe pejorative human qualities to a place.[2] These attitudes inform mainstream western culture that denigrates wetlands and marginalizes wetland communities. Traces of the ancient Greek attitude to wetlands persist today in the association with wetlands of physical and mental illnesses, such as malaria and melancholy, and with monsters and the monstrous. Traces of the ancient Roman attitude to wetlands persist in the association with wetlands of terror, horror, hell, monsters, barbarians and primitives. The history and role of marshes in ancient Roman culture has been researched and traced by Borca (1997, 2000).

Ancient Greek and Roman wetland cultures were syncretized with Judeo-Christian theology by Dante Alighieri in the Italian Middle Ages and by John Milton in the English Renaissance and Reformation to produce the mainstream western cultural denigration of wetlands as evil places and demonization of their inhabitants as monstrous. Also in the seventeenth century, Robert Burton philosophized the melancholy marshes, John Bunyan mapped 'the slough of despond' in the allegorical progress

[2] For further discussion of placism, see Giblett (2018, 12–13, 33).

of Christian and William Dugdale legitimated drainage as divine work by misquoting the biblical creation story. The old English elegiac long poem *Beowulf* had already added a quasi-biblical and Germanic twist to the mix with its marsh monsters and evil dragon. Judeo-Christian theology contributed its own take on watery monsters, with Leviathan (variously a crocodile or a whale) and dragons, culminating in the legend of St George killing a dragon.

All of these facets contributed to the construction of the Christian and mainstream western denigration of wetlands. Instead of the ancient and traditional view of wetland cultures that wetlands are sacred places imbued with spiritual and ceremonial significance, the mainstream western view regarded wetlands as slimy sloughs of despond where monsters such as Satan and dragons live. Mainstream is an appropriate figure for encapsulating western cultures' anti-paludalism, obsessed as they are with streams of flowing water in rivers, creeks and oceanic flows (the Gulf Stream) and their disparagement of stagnant, non-flowing water in some wetlands.

The Gulf Stream off the eastern coast of the United States is vital for life in the region. Wetlands are also vital for life on earth. The leading intergovernmental agency on wetlands states that:

> they are among the world's most productive environments; cradles of biological diversity that provide the water and productivity upon which countless species of plants and animals depend for survival. Wetlands are indispensable for the countless benefits or "ecosystem services" that they provide humanity, ranging from freshwater supply, food and building materials, and biodiversity, to flood control, groundwater recharge, and climate change mitigation. Yet study after study demonstrates that wetland area[s] and [their] quality continue to decline in most regions of the world. As a result, the ecosystem services that wetlands provide to people are compromised. (Ramsar Convention Bureau n.d.)

The Ramsar Convention Bureau website provides a wealth of information, including the global policy context around wetlands. Its 2018 'Global Wetland Outlook' pointed out that wetlands are one of the most rapidly declining ecosystems in the world and addressing 'human dimensions' is at the core of conserving them. Studies of wetland cultures, such as the present volume, more specifically critique denigratory views and destructive practices and promote affirmation and conservation of wetlands locally and transnationally.

1.3 Wetlands are Non-binary, Queer Bodies of Water

Wetlands punch way above their weight as they cover about 6% of the surface of the earth. They are in the minority and they are an oppressed minority and a marginalized community of plants and animals (humans and more-than-humans). This radical political slant on the context for wetland conservation has not been taken up in global policy on wetlands, hence the need for the present volume and its dedication in these terms. Their identity and very existence are under threat from the majority of lands (drylands) and waters (rivers, seas and oceans) and from centralized communities of plants and animals (humans and more-than-humans), including towns and cities. With more humans now living in cities, and with cities generally inimical to wetlands, these threats are exacerbated. Wetlands once were where cities are now; where wetlands are now on the margins of cities, the suburbs will be soon. Where a city now is, a swamp once was.

Such is the case with the city of Cairns in far north Queensland that famously arose 'out of the swamp' and went from 'a struggling village in the mangroves to a thriving tropical city,' as Curzon-Siggers (1996) puts it in his book of this title and subtitle. A similar history could be probably be related with other cities in Queensland, such as the coastal city of Mackay set in mangroves, though there is no mention of this by Emily Emma Brown (2023). A similar history can be related with many cities elsewhere in Australia set in or beside wetlands (such as Perth, Sydney and Melbourne) and around the world, including all the iconic cities of modernity founded in ancient or modern times (London, Paris, Venice, St Petersburg, Berlin, Hamburg, Boston, New Orleans, New York, Toronto, Washington and Chicago).[3] None of these cities is recognized as a Global Wetlands City under the UN's Convention on Wetlands. In November

[3] For further discussion of cities set in, or beside, swamps, marshes, etc., including Tenochtitlan (the fabled floating city of the Aztecs largely destroyed by the Spanish, but remnants of which survive in Mexico City to this day), and all the iconic cities of modernity founded in ancient or modern times in Europe and North America (London, Paris, Venice, St Petersburg, Berlin, Hamburg, Boston, New Orleans, New York, Toronto, Washington and Chicago), see Giblett (2016). For further discussion of Australian cities in similar settings, such as: Melbourne, a city of ghost swamps, see Giblett (2020b, part I; 2020c, Chap. 8; 2021, Chap. 8); Sydney, a swamp and sandstone city, see Giblett (2021, Chap. 7); and Perth, the black swamp/swan city, see n1 above.

2022, the city of Chibayish in the Iraq marshes (discussed in Chap. 3 of the present volume) was nominated as a Global Wetlands City (Lonergan and Al-Asadi 2024, 168).

Writing a word for wetlands is speaking on behalf of marginalized communities and oppressed minorities. Wetlands cannot speak on their own behalf. Nor can their native inhabitants, such as ducks, a point reinforced in the current campaign in the state of Victoria in Australia by the Greens political party to stop duck-shooting: 'Native ducks can still be shot—will you use your voice to speak for them?' (broadcast email to Victorian Greens, 1 May 2023). Wetlands and their native inhabitants are like children and slaves in the past and present as they need others to speak for them, as they cannot speaks for themselves. Wetland conservationists do so, including in the written word of wetland literature.

Wetland literature is a minor literature written in a major language that disrupts from within majority languages and mainstream anti-paludal culture. Wetland literature is akin to what Gilles Deleuze and Felix Guattari called 'the minor literature' of the work of the Czech writer Franz Kafka living in Prague and writing in the major language of German who disrupted it and its imperial culture from within (see Giblett 2021a, 85–86). Wetland literature has been marginalized in the canons of major national literatures, even within the opus of canonical writers, such as Henry David Thoreau and his writings on swamps and Aldo Leopold and his writings on marshes, both of which were neglected or overlooked for a long time. Wetland literature is doubly marginalized within the marginalized literature of 'nature writing.' This radical political slant on the context for wetland conservation has not yet been taken up in global policy on wetlands.

Marginalized communities are usually regarded as human communities composed of gender and sexually diverse, gay, lesbian, bi-sexual, asexual, queer, non-binary, intersex and transgender + peoples forming the rainbow coalition. Wetlands are also non-binary: neither land, nor water, but both; neither solid, nor liquid, but both. Their identity is fluid, transgeneric and indeterminate. Wetlands are queer bodies of water. People of colour are also a marginalized community. Wetlands are places of colour and marginalized communities of plants and animals. More specifically, wetlands in mainstream western culture and its colour-coding are black with all its incipient racist associations.[4] Wetland figured as 'black waters' would be a convenient shorthand label to encapsulate more than two

[4] For further discussion of black, see Giblett (2013, Chap. 14).

millennia of mainstream western denigration and destruction of wetlands (see Giblett 2021b). Wetlands are black waters, distinguished from white water and discriminated against accordingly in word and deed by whites (see Giblett 2007, 2009, Chap. 11).

A prominent instance of figuring and denigrating a wetland as black water can be found in Wilkie Collins' *The Woman in White* of 1859–60 set mainly in Blackwater Park with its Blackwater Lake.[5] In the early twentieth century, black waters, or variations on it, were used many times to describe the swamp of the Bora in C. S. Forester's classic adventure and love story, *The African Queen* first published in 1935 (Forester 1956, 109, 110, 116, 118, 136, 137). It was the basis for the highest grossing film of the 1950s starring Humphrey Bogart and Katherine Hepburn and directed by John Huston. Books and films contributed to the denigration of wetlands.

In the late twentieth century 'black water' was applied to a polluted wetland, or watery wasteland, in Joyce Carol Oates' 1994 novel *Black Water* based on the infamous drowning in 1969 of Mary Jo Kopechne in Senator Edward Kennedy's car on Chappaquiddick Island in Massachusetts. Oates also meditated on other colours of water and contrasted black water with the blue water of swimming pools.[6] More recently, 'blackwater events' occur when flooding washes organic material into waterways generating toxic algae blooms that kill thousands of fish and distresses local fishers and farmers and all conservationists. Black water has shifted from denigrating living wetlands, through describing polluted wet-wastelands, to critiquing dead waterways.

Marginalized communities of peoples act and agitate in solidarity with each other and their allies in common cause—social, cultural and political. As such, they are anthropocentric and do not consider and act in solidarity with other diverse, marginalized communities, such as wetlands with their plants and animals, humans and more-than-humans. Wetland communities could be a part of a much broader, more inclusive rainbow coalition that values the green as much as the other colours. I am yet to find a section in any bookshop devoted to wetland literature that sits side-by-side with the Aboriginal, indigenous, gay, lesbian, queer, bi-sexual, non-binary and transgender + sections.

Wetlands are marginalized communities partly because of their identity. Wetlands have been gendered, or more precisely feminized, in the

[5] For further discussion of *The Woman in White*, see Giblett (1996, 4–8).

[6] For further discussion of *Black Water*, see Giblett (2007, 2009, Chap. 11).

gendered construction of reality. Wetlands are feminized for either good or ill: ill in misogynist patriarchy as places of disease, death, horror, the monstrous and tombs of dead matter and good in matrifocal ecofeminism, especially French feminism, as places of life, death, health, the maternal and wombs of new life.[7] The latter deconstructs the binaries of life and death, womb and tomb, monstrous and maternal. Wetlands are places of both.

Dredging, draining and reclaiming wetlands is a quintessentially masculinist thing to do. It was certainly the kind of war the fascist Benito Mussolini liked to fight and win in Italy with t the Pontine Marshes in the 1930s. Saddam Hussein did likewise with the marshlands of southern Iraq in the 1990s. Floods in wetland cities, such as Venice, New Orleans and New York, suggest that in the longer term the war was never won, never could be and at best only a truce could be achieved in the short term with the will to fill or drain wetlands and under the drive to dredge, drain and reclaim.

1.4 Wetlands in Paintings, Literature and Sculpture

Part of the problem with wetlands for mainstream western culture lies in their aesthetic qualities, or the lack of them. Wetlands often look ugly, smell bad, feel slimy, sound eerie and are uncanny. They are the uncanny place *par excellence*. As a result, they have been denigrated, or they have been aestheticized within the conventions of European landscape aesthetics in the drive to fill their lack of aesthetic qualities and remedy their sensory ills.[8] This aestheticization occurs not only in landscape painting and writing, but also in landscape architecture by transforming wetlands into lakes in parks and gardens.[9] By contrast, Aboriginal wetland peoples have used and appreciated wetlands for tens of thousands of years, including in their art.

[7] For further discussion of all these aspects, see Giblett (1996).

[8] For further discussion of aesthetics and wetlands, including their uncanniness, see Giblett (1996, part I), and wetlands in Australian (including Aboriginal) painting and photography, see Giblett (2020a). For the uncanny in philosophy, psychoanalysis, culture, nature, cities, technology and literature, see Giblett (2019a).

[9] Many instances spring to mind, such as Central Park in New York (see Giblett 2016, 149–152) and the Royal Botanic Gardens in Melbourne (see Giblett 2020b, Chap. 7).

Fig. 1.1 Thomas Thompson *Lake Wendouree, Ballarat* Circa 1875. Oil on canvas. Purchased and conserved with funds from the Joe White Bequest and the Hilton White Bequest, 2006. The Collection of the Art Gallery of Ballarat. Courtesy of the Art Gallery of Ballarat. Reproduced with kind permission

Two paintings of Lake Wendouree in the regional city of Ballarat in western Victoria in Australia displayed permanently in the Art Gallery of Ballarat demonstrate and illustrate this difference. They are hung one above the other in a stroke of curatorial genius, stark cross-cultural contrast and decolonizing context. One was painted by Thomas Thompson in 1875 (Fig. 1.1) and shows the lake with yachts and rowboats on the water, while white people in their Sunday best promenade and picnic on the grassed banks of the lake. Ballarat was a rich city at the time as a result of the Victorian gold rush and Lake Wendouree was the playground and recreational site for the conspicuous display of wealth.

The other painting was painted by Aunty Marlene Gilson in 2016 (Fig. 1.2) and shows the same lake at some point in the past with Aboriginal people in canoes spearing fish and with black swans and other birds on the water, while other Aboriginal people are netting fish on the edge of the water and camping there. Thompson shows no native animals on the

Fig. 1.2 Marlene Gilson *Black Swamp – Lake Wendouree* 2016. Synthetic polymer paint on linen. Purchased with funds from the Art Gallery of Ballarat Association, 2018. The Collection of the Art Gallery of Ballarat. Copyright © Marlene Gilson. Courtesy of the Art Gallery of Ballarat and Martin Browne Contemporary. Reproduced with kind permission

water, on the land, or in the air. They are absent, or have been absented. In Gilson's painting, Bunjil, the all-creator depicted as an eagle, flies above whereas Waa, the trickster, culture hero and ancestral being portrayed as a crow, roosts in a tree.

The juxtaposition of these two paintings highlights how the two different peoples use the land and water: one for recreation and the display of wealth derived from mining and commerce, with the scene as the passive backdrop for them and with little trace of being Australian Aboriginal Country; the other for sustaining their lives and livelihood with the place and its beings as active participants with them for doing so. In one scene, the land, water and air are dead, the surface for recreation; in the other, the land, water and air are living, the depths and heights for sustenance.

Peoples, places and practices (material and spiritual) are lovingly and inextricably intertwined in Aboriginal Country. Here there is no 'landscape' to be visually surveyed, culturally constructed, or greedily consumed; no dead surface of the land, but living depths of the wetlands. Here there is a loving union of people and place, culture and country. Here people, place and practices are not separate, as they are for colonial cultures where places and landscapes are the stage and background for the performance of the engineering and architectural practices of draining or filling swamps, landscaping parks and gardens out of them, pursuing recreational activities and displaying wealth and power in them, and for establishing farms and building homesteads, towns and cities in them, or beside them, or where they once were. Places and landscapes for colonial cultures are also the stage and background for the taking place of agricultural and pastoralist practices by cavalier colonizers—the cavalry—riding roughshod (literally and figuratively) and droving sheep and cattle—the infantry—over Aboriginal places, people, law and traditions (as we will see in the second chapter of the present volume).[10]

The stories of ancestral beings creating and recreating Aboriginal Country in the Dreaming now-time, or 'Everywhen' (McGrath et al. 2023), often include the Rainbow Serpent who created and lives in water bodies (aquifers, rivers, lakes, swamps, etc.) and who embodies water. The Yubiwara people of the Mackay region in north Queensland are no exception as they tell stories of the Rainbow Serpent who is, according to Brown (2023, 10), 'a symbol of fertility, creation and abundance [and] a source of power and wisdom' [who is] revered as a spiritual guide and protector.' These facets have been foregrounded recently by Aboriginal and non-Aboriginal Christians and theologians who have reconstructed the Rainbow Serpent into 'the Rainbow Spirit' in order to defuse the negative biblical connotations of Satan the serpent in the book of *Genesis*, and of water beings in the Bible more generally, as evil. Later literature, such as the long old English elegiac poem *Beowulf* and John Milton's longer modern English epic poem *Paradise Lost,* perpetuates the theology of marsh monsters and swamp serpents as evil beings, whereas they should be regarded as figures that embody and express earthly sacrality of spirituality and materiality, like the Rainbow Serpent/Spirit and as they do in Alexis

[10] For further discussion of: landscape, see Giblett (2011, part II); Aboriginal country, see Webb (1996) and Giblett (2011, Chap. 11); and Australian pastoralism, see Giblett (2011, Chap. 10; 2019b, Chap. 6).

Wright's *Carpentaria* and *The Swan Book* (discussed in the second chapter of the present volume).[11]

A wetland restoration project in south-eastern Australia also acknowledges and promotes the spirituality and materiality of wetlands for Aboriginal people. It is an exemplary collaboration between the local community, conservationists, governments, traditional owners and artists. The Winton wetlands are located north-east of Melbourne between the towns of Benalla and Wangaratta in north-eastern Victoria. The Winton wetlands project is 'a project of national scientific, cultural and environmental significance, [and is purportedly] the largest wetland restoration project in the southern hemisphere' ('Project Overview,' Winton Wetlands Committee of Management n.d.).

The site lies within the traditional lands of the Yorta Yorta people whose 'sovereignty was never ceded' and who play a significant role supporting the restoration project and its programmes. These include a sculpture walk called 'Lotjpatj Natjan Danak,':

> a creation of cultural works that represent Ancestors; totems and special places; spirituality; ceremonies and cultural practices; histories and knowledge systems; ecology and seasons; plants; insects; birds and animals; rivers; creeks; sky stories and ancient creations. These works have been created by fifteen Yorta Yorta artists to share stories of their living culture. ('Sculpture Walk,' Winton Wetlands Committee of Management n.d.)

This descriptor ticks all the boxes for showcasing a pro-paludal, traditional culture on a site and in a place that 'always was, always will be' Aboriginal Country. In late 2023, three exhibitions of wetland art were held in three different galleries in the region (Winton Wetlands Committee of Management 2023).

The mainstream anti-paludal and colonial history of the area is also acknowledged and critiqued, beginning with the account of:

> the flooding of Winton (Mokoan) Swamp in 1971 to create Lake Mokoan [, …] a large water storage that irrigated thousands of hectares of agriculture […] At the same time the new reservoir inundated around 7,000 Ha of

[11] For further discussion of: *Beowulf*, see Giblett (2018, Chaps. 2 and 3; 2021, Chap. 1); Leviathan, dragons and other watery and marshy monsters in the Bible, see Giblett (2018, Chap. 4); and the Rainbow Serpent and the Rainbow Spirit, see Giblett (2018, Chap. 9; 2020d).

> agricultural land, ancient forests and beautiful and unique wetlands. The flooding killed around 200,000 river red gums, including many Aboriginal scar trees that are still standing today, having re-emerged from the lake as dead stags, but still retaining their scarring and their stories. ('Lake Mokoan and its Decommissioning,' Winton Wetlands Committee of Management n.d.)

The skeletal trees are memorials celebrating the life and mourning the loss of the original wetland, as well as a stark reminder of what has been lost and what cannot be regrown in under 200 years. These ghost gums tell their own stories to those who listen and hear the sad and sorry story of the place in the past, present and the foreseeable future. The cover image of *Wetland Cultures,* 'A Dawn Sunstar in the Winton Wetlands,' depicts the ghost gums and presents a glimmer of hope for it and all wetlands.

This image is a fitting emblem for having hope in wetlands and waking to nature that Thoreau expressed in 'Walking' and in *Walden*. 'Hope and the future' for Thoreau (1982, 611) writing in 'Walking' 'are not in lawns and cultivated fields, not in towns and cities, but in the impervious and quaking swamps.' Thoreau is 'a swamp lover' (Thoreau 2017). He is also a dawn waker. Thoreau (1982, 572) wrote at the end of *Walden,* 'only that day dawns to which we are awake. There is more day to dawn. The sun is but a morning star.' Waking to wetlands and having hope in them, as well as waking to nature in general and 'staying woke,' are urgent political and cultural tasks. Thoreau showed the way in all four paths.[12]

The Winton Wetlands project is not only exemplary for showcasing pro-paludal, traditional culture and acknowledging mainstream colonial and anti-paludal culture. It is also unusual as the wetland concerned was flooded when it was impounded as a reservoir and later restored when the impoundment was breached. Rather than being drained or filled to create agricultural or urban land (as is usually the case with mainstream colonial and anti-paludal culture), in this case wetlands and agricultural land were flooded and destroyed in 1971 to create a reservoir to irrigate agricultural land elsewhere. Nearly 40 years later, in 2010, the reservoir was 'decommissioned,' or the impoundment breached, to recreate a wetland for pro-paludal, environmental and traditional cultural uses. This case is also

[12] For further discussion of Thoreau, loving swamps, having hope in them, waking and 'staying woke,' see Giblett (2024).

unusual as a wetland reclaimed the flooded land, unlike the wetland being reclaimed for dry land (as is also usually the case with mainstream colonial and anti-paludal culture).

Both draining wetlands to create dry land and impounding wetlands to create reservoirs are acts of colonization. Wetlands were colonized in Australia and elsewhere in the world, such as in Canada (as we will see in Chaps. 8 and 9 of the present volume). Colonized wetlands need to be decolonized. As Frantz Fanon, the pioneer theorist of decolonization wrote in 1961:

> hostile nature, obstinate and fundamentally rebellious, is in fact represented in the colonies by the bush, by mosquitoes [from swamps], natives and fever [from mosquito bites], and colonization is a success when all this indocile nature has finally been tamed. Railways across the bush, *the draining of swamps* and a native population which is non-existent politically and economically are in fact *one and the same thing*. (Fanon 1967, 201; my emphases)

Draining wetlands to produce dry land or impounding them to create reservoirs are not an add-on to colonization, but part and parcel of it. Decolonization involves decolonizing wetlands by critiquing these aspects of colonization and by promoting pro-paludal, environmental and traditional cultural values of wetlands, such as the Winton Wetlands Project does on both counts, as do the juxtaposition of the Thompson and Gilson paintings in the Art Gallery of Ballarat. Other Aboriginal art and some recent non-Aboriginal artworks do so too by presenting the cultural values of wetlands, their plants and animals for sustaining material and spiritual life. Long live wetlands material and spiritual!

These values are foregrounded and promoted in two classics of Australian children's literature that tell the stories of a wetland and its inhabitants, both human and more-than-human. Waterbirds feature in two of Colin Thiele's books: *Storm Boy* first published 60 years ago that tells the story of a special bond between a boy and a pelican (Thiele 2023); and *Swan Song* that tells the story of a special bond between a boy, two pelicans and a black swan (Thiele 2002). *Storm Boy* also features a special relationship between the boy and an Aboriginal man who is an outsider to mainstream society. It has been made into a film on two occasions. Both books are set in the Coorong wetlands where the Murray River meets the sea in South Australia. This is an internationally important wetland nominated and protected under the Ramsar Convention. Both books tell the story of the place and its peoples and their relationships with birds.

In *Storm Boy* the boy learns from the Aboriginal man who:

> could point out fish in the water and birds in the sky [...] He knew all the signs of the wind and weather in the clouds and the sea. And he could read all the strange writing on the sandhills and beaches—the scribbly stories made by [all the animals that called this place home too]. Before long Storm Boy had learnt enough to fill a hundred books. (Thiele 2023, 9)

Like all good children's literature, adult can appreciate and learn from it too, and learn from Aboriginal people.

Storm Boy evokes in a typical piece of nature writing the sight, sounds and antics of birds bathing in the Coorong wetlands in the early morning light and Storm Boy's response of feeling 'the excitement and wonder of it' (Thiele 2023, 16–17). This scene is later contrasted starkly with the sounds and smells of the shooting when:

> all along the Coorong the shooters went, the blast of their guns echoing up and down the water, and the stench of their gunpowder hanging on the still air like a black fog of rotting smoke. The mornings were filled with the cries and screams of birds. (Thiele 2023, 67–68)

The black fog and bad smells of martial mal-aria ('bad air') are produced by the weapons wielded in the war against wetlands' more-than-human inhabitants, not by the wetland itself and its marsh miasma. These armaments of violence also inflict pain on its innocent and unsuspecting inhabitants. With such harrowing descriptions of shooting, the cruel behaviour of the shooters (Thiele 2023, 18–19, 68–69), the suffering of the birds, as well as conveying the horror and terror of war, and with its tragic, but life-affirming ending, *Storm Boy* is probably the most powerful tract ever produced against waterbird shooting in Australia. All the politicians in Victoria currently considering the future of duck-shooting in the state should read it. It only takes half-an-hour.

Swan Song is an equally powerful tract against messy campers, or 'yahoos' (14), who leave their rubbish behind in environmentally sensitive areas, such as the Coorong wetlands, and against environmental vandals in vehicles doing damage in them too. These 'petrol heads' (16) and 'hoons' (17) driving 'dune-buggies or four-wheel drive jalopies with hyped up motors' described as 'demented metal hornets' (15) are a danger to themselves and others, and inflict injury on both, as well as damage the place.

In *Swan Song*, like *Storm Boy,* Thiele also writes a word for wetlands and their waterbird inhabitants. Mitch, or 'the bird boy,' the main character of *Swan Song*, rescues two pelican chicks and a black swan egg from certain death from exposure or vandalism. He raises all three. He feeds the chicks fish he catches in the Coorong or gets from a local fisherman. With the help of his mother, he incubates the egg until it hatches into an ugly cygnet. The black swan grows into 'one of the most beautiful birds in the world' (72) and 'a magic bird' (190). They call the black swan 'Bugle' because of its musical cry, 'that lovely fluting sound' (63). Bugle becomes the boy's best friend and companion, 'just like the story of the snow goose,' one of the locals says (84), a reference to Paul Gallico's war-time story of this title. With *Storm Boy* and *Swan Song,* Thiele is a kind of Australian Gallico.

Like *Storm Boy, Swan Song* evokes the sight of the Coorong wetlands and describes Mitch's response to seeing them from a distance framed by windows, 'like the loveliest paintings' (9):

> the water in the Coorong was so calm that it lay before him like a huge mirror. At other times, when the breeze blew like a gentle breath and rucked up tiny wavelets, the mirror turned into ripple glass, brightened by sunlight or dulled by passing clouds [...] Often it [the scene] was made even more magical by great flocks of waterbirds that lived on the Coorong or came to visit it from distant places. (10)

The migratory birds that come to the Coorong include waders, such as sandpipers. Mitch's father, the local wildlife ranger, explains that they are:

> some of the best long-distance flyers in the world. They would have come down from the Arctic last October—a little trip of 10,000 kilometres or so across the globe. They'll feed up here until April and then fly back again. And so it goes on year after year. Mitch gazed at the birds in wonder. 'It's a bit of a miracle, to fly across the world like that, just to come here every year?' 'It's magic,' his father answered. 'And there are hundreds of miracles like that going on around us every day.' (22)

Wetlands are habitat for trans-equatorial migratory waterbirds, including waders, on the various flyways of the world (Darby 2020).

The inhabitants of the Coorong wetlands not only vary annually, but also the wetlands themselves change seasonally:

> the Coorong was never quite the same from season to season or from year to year. It changed from winter to summer, and sometimes it changed because something unusual happened. If the River Murray came down in flood[,] the huge barrages at the river mouth had to be opened and torrents of water thundered out into the nearby reaches [of the ocean]. If there was a very high tide that surged in from the sea, or very strong winds blowing onshore, the water in the Coorong was likely to rise. And in the heat of summer much of the far end of the south lagoon evaporated, leaving oozy mudflats and rotting plants that sent out an awful stench. (30–31)

These and other wetlands can look picturesque when viewed from a distance and framed by windows, but they can look and smell bad in closer proximity with their 'stinking ooze' (141). They can also be dangerous places, such as when a storm hits (149–155), and they can have 'beautiful, but dangerous' inhabitants, such as snakes (137–138). *Swan Song* also highlights the issue of fluctuating water levels in artificially maintained and managed wetlands, such as the Coorong.

For a long time the Coorong wetlands have been, and still are today, a site of struggle between fishers, shooters and conservationists, not least around water flows to make the wetlands viable for all concerned, including humans and more-than-humans, such as birds and fish (Horn 2023; O'Gorman 2021, Chap. 5). *Storm Boy* and *Swan Song* can help all players and interested parties, or 'stakeholders,' to respect all the creatures of the Coorong wetlands and to conserve the place itself and its wetland values.

Fluctuating water levels are a feature of all wetlands, such as the artificially maintained and managed Coorong wetlands. They are also a feature of those that are not artificially maintained and managed, such as remote and unmodified wetlands like Lake Eyre-Kati Thanda. This is another iconic South Australian wetland whose basin stretches into Queensland (as discussed in the second chapter of the present volume), as well as into New South Wales and the Northern Territory. It features prominently in the paintings of the great Australian artist, John Olsen (1928–2023). He is justly famous for his paintings of Lake Eyre-Kati Thanda:

> that vast saltpan in arid inland Australia, so often a stark dry lake, full for the second year in a row [2019]. Artist John Olsen has painted the lake many

> times: its extreme isolation and emptiness, as well as its spectacular changeable nature [that] were a fertile inspiration. (Artravelife n.d.)

When water is present, Lake Eyre-Kati Thanda is also occasionally a fertile place with 'hundreds of thousands of banded stilts and pelicans flocking to [feed and] breed in the middle of the desert' (Artravelife n.d.). Olsen called one of these paintings 'the Desert Sea.' This is an ironic or mocking reference to those early European explorers who went on expeditions of discovery in search of 'the inland sea' that they thought must exist in the interior of such a vast continent as Australia. In a way, they were right, but only when it rained heavily and the lake extended from South Australia back up its basin into New South Wales, the Northern Territory and Queensland.

Olsen is a kind of latter day St John the Baptist who was a voice crying in the wilderness preparing the way of the lord of viewing the inland, including wetlands, as sacred and bringing back a message of salvation to benighted urban dwellers. To do so, Olsen took the aerial, God's eye view in his Lake Eyre-Kati Thanda paintings, often quite literally from a plane too, with the result that:

> the artist hovers over his subject, looking down at it from the air, that vast shallow lake with its rippled dunes and inlets and grooves of salt, its patterns and colours and textures and moods changing with the light. (Artravelife n.d.)

Olsen is often credited with opening the eyes of non-indigenous Australians and enabling them/us to see the land and water, such as Lake Eyre-Kati Thanda, in new ways with greater appreciation for their unique environmental features and exquisite aesthetic qualities. Whether this shift includes seeing other wetlands elsewhere in Australia in the same way and with greater appreciation is another question. The vast, iconic wetlands in remote areas of the isolated inland or exotic north of Australia are usually more highly regarded (especially if they are a tourist attraction), than the local, mundane swamp in the south down the end of the street, such as the wetlands of Toondah Harbour in the city of Brisbane in Queensland (as we will see in the second chapter).[13] Both sorts of wetlands are important for humans and more-than-humans.

[13] Also such as Forrestdale Lake in Perth at the end of the street where I lived for 28 years; see Giblett (2006, 2013, part 1).

1.5 The Swampy Pathway Afoot

Chapter 2 of *Wetland Cultures* develops the theme of pro-paludal, traditional cultures by engaging in mutually respectful, cross-cultural dialogue for caring for Country by indigenous and non-indigenous peoples. There will be no equitable reconciliation in Australia between its indigenous owners and everyone else unless and until truth-telling takes place with the stories of people and place as outlined in *The Uluru Statement from the Heart*. Chapter 2 of *Wetland Cultures* discusses wetland cultures in Queensland, beginning with the account of James Morrill, a shipwrecked sailor, who lived and thrived among Aboriginals in the swamps of north Queensland for 17 years from 1846. His account of his swampy sojourn was published in 1864.

During the same period the colonial landscape painter Conrad Martens completed a conventional painting of a swamp on a pastoral 'run' (or station) on the Darling Downs in Queensland. Another painting by him depicts the pastoral landscape on the same 'run' with Aboriginal figures in the foreground. His paintings are discussed in the second chapter of *Wetland Cultures* and contrasted with two paintings of swamps by Mavis Ngallametta of the Kugu-Uwanh people, Putch clan, in the Aurukun area in far north Queensland. This is an important area for wetlands and local Aboriginal people as documented by the Queensland Museum (2022, 351–413). Local Aboriginal people point out that for them 'wetlands mean life' (Queensland Museum 2022, 412–413). Her traditional and contemporary swamp paintings are exquisite and intimate artworks that embrace both eras. Rather than swamps with their negative connotations, the wetlands they depict are sometimes described as lagoons.

Chapter 2 goes on to discuss the refraction of Queensland swamps in mainstream colonial culture from the 1930s to the 1980s with a shift from popular nature writing by 'Mopoke,' to environmental science by J. W. Winter, both of whom ignore or overlook Aboriginal wetland cultures. An exception to these writers during this period and a dissenter to mainstream western wetland culture is Judith Wright (1915–2000), the noted writer and 'the patron saint of Australian conservation,' who lived much of her life in Queensland. During the 1970s and 1980s she critiqued the draining of wetlands and the very word 'landscape,' celebrated the life of a swamp plant (albeit a weed) and campaigned to 'Save the Great Barrier Reef' and for Aboriginal land rights. A contemporary dissenter is Alexis Wright, a member of the Waanyi nation of the southern highlands of the

Gulf of Carpentaria in Queensland and the author of *Carpentaria*, the winner of the prestigious Miles Franklin Literary Award in 2007, and *The Swan Book*. The work of both Wrights is discussed in the second chapter of *Wetland Cultures* where two Wrights make it right, or write it right.

Chapter 2 also discusses the refraction of Queensland swamps in contemporary mainstream culture. The 'lavishly illustrated' book, *Wetlands of Queensland*, with over 500 photographs was published by the Queensland Museum in 2022. This weighty tome of over 400 pages includes brief accounts of Aboriginal peoples' traditional attitudes and behaviour towards specific wetlands in Queensland. These are interspersed between lengthy and detailed discussions of all the scientific aspects of the wetlands of Queensland, but very few of the colonial impacts, such as canal estates, and none of the current development threats, such as to the wetlands in Toondah Harbour, are discussed.

The only comparable book devoted to wetlands in the other Australian states seems to be the book about Western Australian wetlands that Hugh Webb and I co-edited and largely co-authored in the 1990s (Giblett and Webb 1996). It is a pioneering work in wetland cultural studies and the transdisciplinary environmental humanities, along with *Postmodern Wetlands* published in the same year (Giblett 1996). The other Australian states seem to have no comparable book about their wetlands. *Western Australian Wetlands* includes Hugh Webb's classic account of Aboriginal Country, including Aboriginal peoples' traditional attitudes and behaviour towards wetlands in Western Australia (Webb 1996).

Unlike *Wetlands of Queensland*, *Wetland Cultures* discusses early colonial accounts of wetlands (such as Morrill's), early natural history (such as 'Mopoke'), early environmental science (such as Winter), the depiction of wetlands in painting (such as Martens and Ngallametta), the refraction of wetlands in writing (such as Judith and Alexis Wright) and the history of Queensland cities set in or besides wetlands (such as Cairns and Mackay as we saw earlier in this introduction). *Wetland Cultures* follows in the swampy footsteps of *Western Australian Wetlands* that discusses early colonial accounts of wetlands, natural history, environmental science, the refraction of wetlands in writing (such as Sally Morgan, Mudrooroo and Mary Durack), and the history of the city of Perth set among wetlands. *Wetland Cultures* also follows in the swampy footsteps of the recent discussion of the depiction of Australian wetlands in painting and environmental artwork (Giblett 2020a, 2021, Chap. 6; 2023, Chaps. 3 and 7).

The wetlands of Queensland play a vital role in Alexis Wright's two novels, *Carpentaria* (2006) and *The Swan Book* (2013). They are not portrayed as pristine places in the present, as is the case of *Carpentaria,* nor in the future, as is the case of *The Swan Book*, but are still revered as traditional sacred spaces created by ancestral beings in the past and recreated in the present. Both aspects lend themselves to expression in magic realism with a uniquely Aboriginal inflection. *Carpentaria* begins by relating the creation story of 'the serpent' who now lives underground in the aquifers. Like Mavis Ngallametta's paintings of lagoons, 'the lagoon' figures in *Carpentaria,* not only as a secret fishing hole and a sacred place, but also as a crime scene. The Gulf of Carpentaria is not only Aboriginal Country, but also mining country, as are many places in Queensland. Three young Aboriginal industrial saboteurs unleash the monstrous serpent of fire against the mine, the miners and their monstrous mining machines and enact a terrible revenge on behalf of the ancestral creative water serpent.

The Swan Book begins in a backwater swamp with its polluted water located near the coast of north Queensland. It is also a graveyard for ghost gums and ghost ships. This is the homeland of the swamp people, including a mute girl who has a special relationship with black swans. She is abducted by a powerful politician and taken to the city of ghost swamps in the south.[14] He orders the evacuation and dredging of the swamp. After he is assassinated, she returns to the ghost swamp in the north. Whereas *Carpentaria* begins with the Rainbow Serpent, *The Swan Book* ends with the Rainbow Serpent. Both books are written in the style and tradition of magic realism, beginning with Gabriel Garcia Marquez's *One Hundred Years of Solitude* and continuing with Salman Rushdie's *Midnight's Children,* among many other notable examples, such as the writing of Angela Carter. *Carpentaria and The Swan Book* are sort of Aboriginal equivalents.

Aunty Marlene Gilson's depiction of Aboriginal people in canoes spearing fish is remarkably similar to Wilfred Thesiger's photograph of 1958 showing Marsh Arabs spearing fish from canoes that now graces the cover of the Penguin Classics edition of his book, *The Marsh Arabs* first published in 1964 (Thesiger 2007). Thesiger admired and was envious of the skilful balance of the Marsh Arabs who stood in the front of their canoes and speared fish. Chapter 3 of *Wetland Cultures* discusses Thesiger's

[14] Melbourne, not named as such in *The Swan Book*. For Melbourne as a city of ghost swamps, see Giblett (2020a, part I).

account of his sojourn among the Marsh Arabs in the 1950s and their paludiculture in the Mesopotamian Marshes in southern Iraq, once the largest wetland system on the planet. Gavin Maxwell joined Thesiger in 1956 in Iraq and wrote his own book, *People of the Reeds* first published in 1957, about his sojourn among these people before Thesiger wrote and published his (Maxwell 1966). Gavin Young started visiting the Marshes in the 1950s and in 1977 first published his account, *Return to the Marshes: Life with the Marsh Arabs of Iraq* (2009). The accounts of the two Gavins are also discussed in Chap. 3.

In the 1990s, the genocidal, anti-paludicultural and aquaterracidal regime of Saddam Hussein drained large sections of the Marshes and dispossessed the Marsh Arabs. Recent efforts are dedicated to rehabilitating these Marshes and restoring the life-ways of the Marsh Arabs. Jassim Al-Asadi, an irrigation engineer, was jailed and tortured under the regime of Saddam Hussein. He subsequently dedicated his life to the reflooding and restoration of the Marshes. He eventually contributed to the Mesopotamian marshes being declared a UNESCO World Heritage site in 2016. In February 2023, he was kidnapped, tortured again and released after 15 days. In April 2024, Steve Lonergan and he published *The Ghosts of Iraq's Marshes* that tells his story and brings the story of the Marsh Arabs right up to date, including with his insider's view interwoven with the story of the Marshes (Lonergan and Al-Asadi 2024). Chapter 3 discusses their work and the work of Jennifer Pournelle (2003) and Edward Ochsenschlager (2004) on the Marsh Arabs and their Marshes, especially Pournelle's pioneering work foregrounding the role of the marshlands as a cradle of western, urban civilizations and cultures.

The mainstream western denigration of wetlands began arguably with the ancient Greeks and their philosophy of the four elements (earth, air, fire and water), and their related miasmatic theories of disease, especially malaria (literally 'bad air'), the 'humour' of melancholia and the bestiary of monsters. Wetlands are aberrant as they mix earth and water. They are considered on this view as bad for bodily and mental health; they are bad in this life and the next. Homer associated wetlands with the underworld of Hades, or hell, the land of the dead. All of these aspects are discussed in Chap. 3 of *Wetland Cultures.*

The ancient Romans inherited and developed all these aspects, especially with Virgil's view of wetlands and Hades, or hell. Dante took up Virgil's view and Christianized it in *The Inferno* with the souls of the sullen stuck in slime in a circle of hell. John Milton later syncretized Dante's Christianized vision of hell with ancient Greek and Roman views. In

Paradise Lost Milton portrayed Satan as a monstrous swamp serpent engendered in slime. In the legend of St George killing a dragon, the dragon is a marsh monster in one version, a swamp serpent in others, and St George is a canonized Christian knight in all of them. The legend became widespread in Europe and the foundation myth of England enshrined in its flag and exported to its colonies. All of these aspects are discussed in Chap. 4 of *Wetland Cultures*.

Mediaeval Italian culture and English Renaissance culture inherited much from ancient Greek and Roman wetland culture. This is hardly surprising as the Renaissance revived ancient Greek and Roman learning after centuries of domination by Judeo-Christian thinking. Yet the Mediaeval and Renaissance inheritance and perpetuation of ancient Greek and Roman anti-paludal wetland cultures has not often been remarked upon. Denigration and drainage of wetlands is a marker of modernity, but they have their roots in ancient Greece and Rome and their anti-paludal cultures.

Ancient Greek and Roman wetland cultures persisted in mediaeval and early modern times in the English Fens with their denigration along ancient Greek lines as places of miasma, malaria, melancholia and the monstrous, and as the hellish underworld along ancient Roman, mediaeval and Renaissance Christian lines. These aspects, including Robert Burton's philosophizing of melancholic marshes, are also discussed in Chaps. 4 and 5 of the present volume, as are briefly the traditional agricultural, aquacultural and paludicultural practices of livelihood performed in the Fens, all of which were denigrated and drainage implemented. These aspects are discussed in Chaps. 4 and 5. William Dugdale's spurious biblical legitimation for drainage and Charles Kingsley's ambivalent appreciation for the undrained Fens and advocacy for Fen drainage are discussed in several chapters. The history of the draining of the Fens has recently been researched and traced extensively by Ash (2016, 2017). The Fens are also the recent trial site for the restoration of paludiculture (Hillsdon 2020). The association between disease and the miasma arising from stagnant waters persisted into the nineteenth century in England with the Sanitary Movement in cities discussed in conclusion in Chap. 4.

Rather than figuring hell as a swamp, and in increasingly secularized and post-Christian cultures, for later Anglophone writers in the nineteenth and twentieth centuries the swamp is hell as discussed in Chap. 5. It is a place of death and destruction. By contrast, for twenty-first century writers attuned to wetland ecology, such as Delia Owens in her recent novel,

Where the Crawdads Sing, swamps are places of both life and death as discussed in this chapter. The conclusion to Chap. 5 rewrites the anti-paludal sections of Milton's *Paradise Lost* for a pro-paludal conservation counter-theology.

Chapter 6 of *Wetland Cultures* goes on to consider the tracing of traditional and contemporary cultures of the English Fens in politics, painting, poetry, prose and art history. Recent work on the English Fens coming out of art history, landscape aesthetics and landscape architecture in Vittoria di Palma's (2014) *Wasteland* demonstrates the still contentious and problematic positioning of the Fens in English culture and the need for concerted efforts to conserve them. This chapter takes a radical political approach to the Fens and to the enclosure of the Commons (including much of the Fens) into private property critiqued by John Clare, Karl Marx, E. P. Thompson and Raymond Williams. It also takes a cultural, literary and historical approach by considering the depiction of the Fens in painting and the literature of the Fens, beginning with Clare's poetry and prose in the nineteenth century for the Fens and against enclosure. It then proceeds to a consideration of the Fens in later prose, beginning with Dorothy Sayers' classic detective novel *The Nine Tailors* (1934), then to Graham Swift's wet-landmark *Waterland* (first published in 1983; Swift 1984) and his 25th anniversary introduction of 2008 (Swift 2010), Robert Macfarlane's (2008) 'new nature writing' of his essay/story 'Ghost Species,' and culminating recently with Becky Owen-Fisher's poetry (Owen-Fisher and Tearle 2017) and Stella Tillyard's historical romance *The Great Level* (2018).

Chapter 7 of *Wetland Cultures* considers how the ancient, traditional and contemporary culture of Irish bogs is traced in twentieth-century poetry and prose beginning with Seamus Heaney who was a recipient of the Nobel Prize for literature and is the patron saint of bogs. He was following in the boggy footsteps of Lord Dunsany's *The Curse of the Wise Woman* published in the 1930s and H. P. Lovecraft's weird tale 'The Moon Bog' published in the 1920s, both of which are discussed in Chap. 7. In the present century, Kelly Barnhill in *The Girl who Drank the Moon* published in 2017 sings the song of the bog. All of this writing appreciates bogs as active participants with agency, as a pagan, pre-Christian wetlandscape, as well as the counter to Christian theology, to English colonization and to industrialization and modernization. They also feminize bogs as maternal figures and spaces with a fertile womb for the creation of new life. These writers also feminize bogs as maternal figures and spaces with a fertile womb for the creation of new life. A supplement to this chapter of

'Rebirth in a Bog' complements the canon of the bog bible. It is excerpted from my book of detective stories, *Swamp Deaths: Collected Cold Cases and Other Marshy Mysteries* (Giblett 2022). The history of Irish bogs, the use of peat from them and the draining of them has been researched and traced recently by Ruuskanen (2018). Irish bogs are still an important source of peat for heating and cooking in the age of rising energy costs (Carroll and Connolly 2022).

From England, Ireland and Europe more generally the focus of *Wetland Cultures* shifts eastward across the northern Atlantic, also the direction of the journeys of early European explorers and colonists. Canada has approximately a quarter of the world's wetlands (marshes, swamps, bogs, etc.). It is the leading country in the world for wetlands with a larger area of wetlands than any other country. Canada is geographically the wetland country *par excellence*. Canada also probably has the dubious distinction of having lost a greater area of wetlands than any other country. Canada has certainly lost or damaged nearly all of its urban wetlands. Canada has the dubious distinction of probably being culturally the anti-paludal country *par excellence*. Historically Canadian wetlands have been involved in the lives of its peoples from the First Nations who came across the land bridge from Asia through the first French, and later English and Scottish explorers, missionaries and colonists who came by sailing ship from Europe to the later diverse diaspora from many European countries who came by steam ship and aeroplane.[15]

Canada is geographically not only the wetland country par excellence. Colonial Canada is historically the wetlands country *par excellence* as its first European settlements were founded next to marshes. The first French explorers put out the cattle they had brought with them to graze on the marsh hay at Cap Rouge in Quebec. Matthew Hatvany argues that when Jacques Cartier landed here in 1541 marsh hay would also have provided:

> thatch and bedding (both for animals and humans), and eels and waterfowl taken from the marsh [...] In 1608, when Champlain founds Quebec, he specifically mentions cutting the same type of marsh hay for the animals along the Cote-de-Beaupré. (Hatvany cited by Giblett 2014, 24)

[15] For further discussion of English and Scottish wetland cultures in Canada, including in the Anglophone literary canon and recent writing, see Giblett (2014).

French, or Lower, Canada has a marshy beginning, as does English, or Upper, Canada founded in 1793 when Lieutenant Governor John Graves Simcoe established its capital at Toronto in Ontario next to Ashbridge's Bay Marsh later dredged and embanked into a harbour (Giblett 2014, Chap. 7). The modern nation-state of Canada has a marshy beginning. This beginning and heritage spanning more than four and a half centuries should be honoured and respected, as well as Canadian wetlands conserved as sites of cultural and natural heritage and as habitats for plants and animals.

Following in the footsteps of French explorers, wetlands sustained the lives of the first French colonists who crossed the Atlantic to settle in Acadia and become Acadians. Chapter 8 considers how Francophone culture both at home and abroad in Canada used the technology of *aboiteaux* (clapper-valve gate) to regulate water flows and produce output from regularly inundated wetlands. Acadians managed species of native wetland plants to feed their sheep and cattle (not themselves, so they were pastoralist and quasi-paludicultural). This chapter also discusses how traditional Francophone and Acadian culture of wetlands is traced in nineteenth- and twentieth-century literature, beginning with Longfellow's Americanized and saccharine inflection of it in *Evangeline* published in 1847 and concluding with Antonine Maillet's novels published in the 1970s that explicitly counter *Evangeline* and help create a pro-paludal contemporary Canadian wetland culture. Maillet was awarded the prestigious Prix Goncourt for her novel *Pélagie*. She was the first Francophone writer not living in France and the first non-French citizen to be awarded this French equivalent of the Booker prize.

The final chapter of *Wetland Cultures* focuses on the Canadian nature writer Harry Thurston who writes a word for wetlands of the marshes and their native animals in his home provinces of New Brunswick and Nova Scotia. He follows in the muddy footsteps of such antecedents as Henry David Thoreau, the patron saint of swamps, and Aldo Leopold, the patron saint of marshes.[16] Thurston is a kind of Canadian equivalent or counterpart to them, a living legend of Canadian wetlands and a national treasure who should be regarded as an international one too. For more than thirty years Thurston has written and published nature poetry and prose, most

[16]For Thoreau, the patron saint of swamps, see Giblett (1996, ix, 229–239—the first instances in print); 2020c, Chap. 4; 2021, Chap. 9) and Leopold, the writer of a wetlands almanac and the patron saint of marshes, see Giblett (1996, 242–245; 2020c, Chap. 7).

recently in 2022 when he teamed up with photographer Thaddeus Holownia to memorialize the lives and mourn the death of tens of thousands of migratory songbirds incinerated in the flame of a hundred-foot-high flare at a liquefied natural gas plant in New Brunswick in Canada in 2013 (Thurston and Holownia 2022, 5). The Canaport LNG company was fined in 2015 at a rate of less than $100 per bird killed, an obscenely low price to put on the death of a living being. Migratory shorebirds and wader birds also traverse the various flyways of the world linking the feeding and breeding sites of marshes, mudflats and other wetlands. Wetlands are waterbird habitat. Environmental conservation is not just a matter of conserving birds, their habitats and flyways, but also a matter of halting and abolishing the fossil-fuel industry that directly destroy lives (such as in the Canaport case) and indirectly destroys lives, livelihoods and habitats through rising temperatures exacerbating fires and floods, and raising sea levels as a result of greenhouse gas emissions. Environmental conservation is also a matter of developing affirmative wetland cultures with Thurston as one guide for doing so as outlined in this chapter and its conclusion.

In summary, *Wetland Cultures* considers:

- ancient, traditional and contemporary Aboriginal wetland cultures in Queensland and non-Aboriginal responses to them from colonization to the present;
- ancient, traditional and contemporary Arabic wetland cultures in southern Iraq;
- ancient Greek and Roman wetland cultures and their dispersion in mediaeval, Renaissance and modern England, and;
- traditional, mediaeval, modern and contemporary wetland cultures in the English Fens, Irish bogs and Acadian marshes in Canada.

Humans can learn to appreciate wetlands in the present by considering how wetlands and their inhabitants were regarded and treated for good or ill in the past and how they are treated in the present. Traditional cultures have an ongoing and contemporary connection to, and appreciation of, wetlands that should be sustained in the present for the future. The denigration of wetlands in the past and present in mainstream western culture and its ancient Greek and Roman antecedents needs to be critiqued, while the appreciation of wetlands in the past and present in traditional and contemporary cultures needs to be valued and respected. The present volume considers ancient, traditional and contemporary wetland cultures. It is only by not denying or repressing, and learning from, the mistakes of the past of the denigration of wetlands, that wetlands can be conserved in the present for future generations of human and more than human life on earth, indeed, for the life of the earth. Long live wetlands and pro-paludal cultures!

References

AAP. 2019. Budj Bim Indigenous Eel Trap Site Added to World Heritage List. *The Guardian*, July 7. https://www.theguardian.com/australia-news/2019/jul/07/budj-bim-indigenous-eel-trap-site-added-to-world-heritage-list. Accessed 7 July 2020.

Artravelife. n.d. John Olsen – Lake Eyre. http://artravelife.com/blog/files/54a6b464ec43c9343dfe4778a90d7b8d-12.php. Accessed 29 May 2023.

Ash, Eric H. 2016. Reclaiming a New World: Fen Drainage, Improvement, and Projectors in Seventeenth-Century England. *Early Science and Medicine* 21: 445–469.

———. 2017. *The Draining of the Fens: Projectors, Popular Politics, and State Building in Early Modern England.* Baltimore, MD: Johns Hopkins University Press.

Borca, Federico. 1997. *Palus Omni Modo Vitanda:* A Liminal Space in Ancient Roman Culture. *Classical Bulletin* 73 (1): 3–12.

———. 2000. Towns and Marshes in the Ancient World. In *Death and Disease in the Ancient City*, ed. Valerie M. Hope and Eireann Marshall, 74–84. London: Routledge.

Brown, Emily Emma. 2023. *The History of Mackay: A Journey Through History, Culture and Nature.* Self-published.

Cahir, Fred, Ian D. Clark, and Philip A. Clarke. 2018. Conclusion: The Future of Aboriginal Biocultural Knowledge. In *Aboriginal Biocultural Knowledge in South-Eastern Australia: Perspectives of Early Colonists*, ed. Fred Cahir, Ian D. Clark, and Philip A. Clarke, 281–281. Clayton South, Victoria: CSIRO Publishing.

Carroll, Rory and Kate Connolly. 2022. 'Like an Oilwell in your Back Yard': Irish People Turn to Cutting Peat to Save on Energy Bills. *The Guardian*, December 12. https://www.theguardian.com/world/2022/dec/12/like-an-oilwell-in-your-back-yard-irish-turn-to-cutting-peat-to-save-on-energy-bills. Accessed 13 Dec 2022.

Curzon-Siggers, John. 1996. *Out of the Swamp: Cairns: A Struggling Village in the Mangroves to a Thriving Tropical City.* Cairns: James Cook University. Advanced Studies in Regional Queensland History.

Darby, Andrew. 2020. *Flight Lines: Across the Globe on a Journey with the Astonishing Ultramarathon Birds.* Crows Nest, NSW: Allen & Unwin.

Di Palma, Vittoria. 2014. *Wasteland: A History*. New Haven, CT: Yale University Press.

Fanon, Franz. 1967. *The Wretched of the Earth*. Trans. C. Farrington. Harmondsworth: Penguin Books. First published in French in 1961.

Forester, C. S. 1956. *The African Queen*. Harmondsworth: Penguin. First published 1935.

Gearey, Mary, and Rod Giblett. Forthcoming. Wetland Cultures. In *The Routledge Handbook of Wetlands*, ed. Alan Dixon. London: Routledge.

Giblett, Rod. 1996. *Postmodern Wetlands: Culture, History, Ecology*. Edinburgh: Edinburgh University Press.

———. 2006. *Forrestdale: People and Place*. Bassendean: Access Press.

———. 2007. Black and White Water. In *Fresh Water: New Perspectives on Water in Australia*, ed. Emily Potter, Alison Mackinnon, Stephen McKenzie, and Jennifer McKay, 31–43. Carlton: Melbourne University Press.

———. 2009. *Landscapes of Culture and Nature*. Basingstoke: Palgrave Macmillan.

———. 2011. *People and Places of Nature and Culture*. Bristol: Intellect Books.

———. 2013. *Black Swan Lake: Life of a Wetland*. Bristol: Intellect Books.

———. 2014. *Canadian Wetlands: Places and People*. Bristol: Intellect Books.

———. 2016. *Cities and Wetlands: The Return of the Repressed in Nature and Culture*. London: Bloomsbury.

———. 2018. *Environmental Humanities and Theologies: Ecoculture, Literature and the Bible*. London: Routledge.

———. 2019a. *Environmental Humanities and the Uncanny: Ecoculture, Literature and Religion*. London: Routledge.

———. 2019b. *Psychoanalytic Ecology: The Talking Cure for Environmental Illness and Health*. London: Routledge.

———. 2020a. Artist and Swamp: Wetlands in Australian Painting and Photography. In *Australian Wetland Cultures: Swamps and the Environmental Crisis*, ed. John Charles Ryan and Li Chen, 51–70. Lanham, MD: Lexington Books.

———. 2020b. *Modern Melbourne: City and Site of Nature and Culture*. Bristol: Intellect Books.

———. 2020c. *New Lives of the Saints: Twelve Environmental Apostles*. Lanham, MD: Hamilton Books.

———. 2020d. Rainbow Serpent Anthropology, or Rainbow Spirit Theology, or Swamp Serpent Sacrality and Marsh Monster Maternity? In *Australian Wetland Cultures: Swamps and the Environmental Crisis*, ed. John Charles Ryan and Li Chen, 33–50. Lanham, MD: Lexington Books.

———. 2021a. *Black Swan Song: Life and Work of a Wetland Writer*. Lanham, MD: Hamilton Books.

———. 2021b. *Wetlands and Western Cultures: Denigration to Conservation*. Lanham, MD: Lexington Books.

———. 2022. *Swamp Deaths: Collected Cold Cases and Other Marshy Mysteries.* London: Europe Books.

———. 2023. *Middlemarsh: The Hopkins River, Kindred Wetlands and Remarkable People.* London: Transnational Press.

———. 2024. *Waking to Nature: With Thoreau and Benjamin.* New York: Bloomsbury Academic.

Giblett, Rod, and Hugh Webb, eds. 1996. *Western Australian Wetlands: The Kimberley and South-West.* Perth, Western Australia: Black Swan Press/ Wetlands Conservation Society.

Gunditjmara People, with Gib Wettenhall. 2010. *The People of Budj Bim: Engineers of Aquaculture, Builders of Stone House Settlements and Warriors Defending Country.* Heywood, Victoria: em PRESS Publishing for the Gunditj Mirring Traditional Owners Aboriginal Corporation.

Hillsdon, Mark. 2020. Dry Run: The Wet Farming Experiment that could Sow Seeds for Future Crops. *The Guardian*, November 12. https://www.theguardian.com/environment/2020/nov/12/dry-run-the-wet-farming-experiment-that-could-sow-seeds-for-future-crops-aoe. Accessed 13 Nov 2020.

Horn, Caroline. 2023. Coorong Restoration Roadmap Released for Public Comment as Experts Outline Plans for Wetlands Renewal. *ABC News*, May 23. https://www.abc.net.au/news/2023-05-17/roadmap-to-coorong-recovery-released-for-public-comment/102356830. Accessed 1 June 2023.

Langton, Marcia. 2021. *Welcome to Country: A Travel Guide to Indigenous Australia.* 2nd ed. Richmond: Hardie Grant Explore.

Lonergan, Steve, and Jassim Al-Asadi. 2024. *The Ghosts of Iraq's Marshes: A History of Conflict, Tragedy, and Restoration.* Cairo, Egypt and New York: The American University in Cairo Press.

Macfarlane, Robert. 2008. Ghost Species. In *Granta: The Magazine of New Writing 102. The New Nature Writing*, 109–128.

Maxwell, Gavin. 1966. *People of the Reeds.* New York: Pyramid Books. First published 1957.

McGrath, Ann, Laura Rademaker and Jakelin Troy, editors. 2023. *Everywhen: Australia and the Language of Deep History.* Kensington, NSW: UNSW Press.

Milner, Jim and Judith Stuart. 2022. Paludiculture – the Future of Farming on Peat Soils? https://naturalengland.blog.gov.uk/2022/09/30/paludiculture-the-future-of-farming-on-peat-soils/. Accessed 3 May 2023.

Ochsenschlager, Edward. 2004. *Iraq's Marsh Arabs in the Garden of Eden.* Philadelphia, PA: University of Pennsylvania Museum of Archaeology and Anthropology.

O'Gorman, Emily. 2021. *Wetlands in a Dry Land: More-Than-Human-Histories of the Murray-Darling Basin.* Seattle: University of Washington Press.

Owen-Fisher, Becky, and Nick Tearle. 2017. *Standing High out of Shrunken Peat.* N.p.: Compass Publishing.

Pournelle, Jennifer. 2003. Marshland of Cities: Deltaic Landscapes and the Evolution of Early Mesopotamian Civilization. Doctoral Dissertation. University of California, San Diego, Anthropology. https://www.academia.edu/843377/Marshland_of_Cities_Deltaic_Landscapes_and_the_Evolution_of_Early_Mesopotamian_Civilization. Accessed 27 Apr 2023.

Queensland Museum, in partnership with the Queensland Department of Environment and Science. 2022. *Wetlands of Queensland: A Queensland Museum Discovery Guide*. Brisbane: Queensland Museum.

Ramsar Convention Bureau. n.d. The Importance of Wetlands. http://www.ramsar.org/about/the-importance-of-wetlands. Accessed 1 May 2015.

Ruuskanen, Esa. 2016. Valuing Wetlands and Peatlands: Mires in the Natural Resource and Land Use Policies in the Nordic Countries from the Late Eighteenth Century to the Present Date. In *Trading Environments: Frontiers, Commercial Knowledge, and Environmental Transformation, 1750–1990*, ed. Gordon M. Winder and Andreas Dix, 118–132. New York: Routledge.

———. 2018. Encroaching Irish Bogland Frontiers: Science, Policy and Aspirations from the 1770s to the 1840s. In *Histories of Technology, the Environment and Modern Britain*, ed. Jon Agar and Jacob Ward, 22–40. London: University College London Press.

Ryan, John C., and Li Chen. 2020. Swamp-philia and Paludal Heroism: The Passion of Wetland Conservationists in Australia and Elsewhere. In *Australian Wetland Cultures: Swamps and the Environmental Crisis*, ed. John C. Ryan and Li Chen, 195–223. Lanham, MD: Lexington Books.

Sayers, Dorothy. 1934. *The Nine Tailors*. New York: Harcourt.

Swift, Graham. 1984. *Waterland*. London: Picador.

———. 2010. Introduction to the 25th Anniversary Edition. In *Waterland: With a New Introduction by the Author*, v–xi. London: Picador.

Thesiger, Wilfred. 2007. *The Marsh Arabs*. London: Penguin. First published 1964.

Thiele, Colin. 2002. *Swan Song*. South Melbourne: Thomas C. Lothian.

———. 2023. *Storm Boy*. Wahroonga, NSW: New Holland Publishers. First published 1963.

Thoreau, Henry David. 1982. In *The Portable Thoreau*, ed. Carl Bode. New York: Viking Penguin.

———. 2017. The Swamp Lover: An Excerpt from his May 1862 Article 'Walking'. *The Atlantic Monthly*. https://www.theatlantic.com/magazine/archive/2017/11/the-swamp-lover/541356/. Accessed 17 Oct 2023.

Thurston, Harry, and Thaddeus Holownia. 2022. *Icarus, Falling of Birds: A Poem by Harry Thurston, Photography by Thaddeus Holownia*. Jolicure: Anchorage Press.

Tillyard, Stella. 2018. *The Great Level*. London: Chatto & Windus.

Verschuuren, Bas. 2018. Religious and Spiritual Aspects of Wetland Management. In *The Wetland Book*, ed. C.M. Finlayson et al., 1405–1415. Dordrecht: Springer.

Webb, Hugh. 1996. Aboriginal Country: Not a Construction, a Way of Being. In *Western Australian Wetlands: The Kimberley and South-West*, ed. Rod Giblett and Hugh Webb, 61–76. Perth, Western Australia: Black Swan Press/Wetlands Conservation Society.

Wichtmann, Wendelin, Christian Schröder, and Hans Joosten, eds. 2016. *Paludiculture – Productive Use of Wet Peatlands: Climate Protection – Biodiversity – Regional Economic Benefits*. Stuttgart: Schweizerbart Sche Vlgsb.

Winton Wetlands Committee of Management. 2023. Art at the Wetlands: Winton Wetlands Inspire Art 'Extravaganza'. https://wintonwetlands.org.au/art-at-the-wetlands/. Accessed 10 Aug 2023.

———. n.d. Welcome to Winton Wetlands. https://wintonwetlands.org.au. Accessed 12 May 2023.

Wright, Alexis. 2006. *Carpentaria*. Artarmon, N.S.W: Giramondo.

———. 2013. *The Swan Book*. Artarmon, N.S.W: Giramondo.

Young, Gavin. 2009. *Return to the Marshes: Life with the Marsh Arabs of Iraq*. London: Faber. First published 1977.

CHAPTER 2

Wetland Aboriginals in Queensland

Aboriginal peoples in Australia have lived by wetlands, and on their resources, for tens of thousands of years. They are not only the longest surviving cultures on earth, but also the longest surviving wetland cultures. They cultivated lands and waters with aquaculture, paludiculture and pasturalism (as distinct from pastoralism). These cultural practices are demonstrations of their sovereignty (never ceded), vital signs of their civilizations and evidence for 'native title.' Their connection to wetlands is not only ancient, but also traditional and contemporary. It is ongoing today. It is also indigenous, non-western, 'pre-modern' and anticolonial—nowhere less so than in the north-eastern Australian state of Queensland. The present chapter traces the wetland cultures of Aboriginal and colonial peoples in Queensland as outlined in the introduction to the present volume and calls for the conservation of wetlands and promotion of pro-paludal cultures. This call is especially pertinent in the contemporary context of the current campaign to save the Ramsar-listed wetlands of Toondah Harbour and their environs in Queensland from destructive development (Readfern 2023).

In 1846 James Morrill was shipwrecked on the coast of Queensland north of the Burdekin River. He lived among the local Aboriginal people for 17 years before returning to colonial white society, writing his account of his sojourn and publishing it in 1864. The coastal area in which Morrill spent much of his sojourn was 'low and swampy in many places' (Morrill

R. Giblett, *Wetland Cultures*, Literatures, Cultures, and the Environment, https://doi.org/10.1007/978-3-031-57365-1_2

2006, 70). The abundant animal and plant life of the swamps was a reliable food source for the local Aboriginal people. Morrill (2006, 33, 70) learnt how to 'snare ducks, wild turkeys, geese and other wild fowls' 'with snares [...] placed in the thick grass and reeds in the swamps.' Women not only gather food in swamps (55) and dig in the earth for roots (55), but also cultivate roots in swamps. He observed women and very young children 'working in the swamps with their mothers, setting roots' (52). The old women made cakes from the roots gathered from swamps (67). This is 'digging-stick farming' and paludiculture, the cultivation of wetlands.

Land is the site of the clash of cultures, not least with the colonizers over possession and perceptions of lands wet and dry. Morrill (50–51) relates that:

> I told them the white men had come to take their land away. They always understood that might, not right, is the law of the world, but they told me to ask the white man to let them have all the ground to the north of Burdekin, and to let them fish on the rivers; also the low grounds, they live on to get their roots—ground which is no good to white people, near the coast and swampy.

Morrill (71) concludes by emphasizing that 'the low swampy grounds' were 'no good to anybody but themselves.' Aboriginal people understood that white people viewed swamps as no good to them.

2.1 Queensland Wetlands in Paintings

Swamps were good to white people if they were aestheticized in conventional landscape paintings and subsumed and used within a larger pastoral setting, such as in Conrad Martens' painting, *Coochin Run* of 1854, completed during the period of Morrill's swampy sojourn (Fig. 2.1):

The Queensland Art Gallery of Modern Art (n.d.-b) points out that "Coochin' means red in the Jagera language, for the red bill of the black swans that frequented the area.' 'Run' means pastoralist station in Australian English and is a euphemism for land stolen from Aboriginal people, squatted upon to 'run' sheep or cattle and later leased from the colonial state to legitimatise this theft after the fact. The black swan is the endemic and iconic waterbird *par excellence* of Australian wetlands and the avifaunal emblem of the state of Western Australia and its capital city, Perth.

Fig. 2.1 Conrad Martens 1801–78 / *Coochin Run* 1854 / Watercolour on paper / 29.5 x 42.8cm / Purchased 1998 with funds raised through The Conrad Martens Queensland Art Gallery Foundation Appeal and with the assistance of the Queensland Government's special Centenary Fund / Collection: Queensland Art Gallery | Gallery of Modern Art / Photograph: QAGOMA. Reproduced with kind permission

This painting bears an uncanny resemblance to Eugene von Guérard's painting 'Mount William and Part of the Grampians' of the year later with a swamp in the foreground and mountains in the background as in Martens' painting (Fig. 2.2):

Except that there are differences between the two paintings. Von Guérard gives more prominence to the swamp in the foreground with its native animals than Martens does with his swamp. In addition, von Guérard does not subsume the swamp within the pastoralist landscape. Indeed, according to one writer on this painting, 'although this scene was on Mount William Station in Victoria, von Guérard chose to paint it as a primeval landscape.' In other words, although the scene was part and parcel of pastoral land use, von Guérard chose to paint it as a primeval hydrological wetlandscape of marsh and slime in the fore- and mid-ground, and

Fig. 2.2 Eugene von Guérard, *Mount William and part of the Grampians in West Victoria*, 1865. Oil on cardboard, 30.3 × 40.6 cm. National Gallery of Victoria, Melbourne. Collier Bequest 1955 (1562–5). Reproduced with kind permission

as a primeval geological drylandscape of rock and the sublime in the background. The horizontal, supine wetlandscape is countered by the vertical, erect drylandscape in the background in a classic, masculinist gendering of the land. Mountainous clouds surmount the scene and shroud the top of Mt William as an emblem of God's overarching omnipotence, if not beneficence.

The pleasing prospect of Martens's pastoralist landscape is bathed in beneficent, Turneresque sunlight as an emblem of God's creation of light out of darkness and dividing land from the water as related in the biblical book of *Genesis*. Whereas von Guérard's foreground depicts the primeval slime of *Genesis* 1: 1 when the world was wetland, Martens' foreground depicts God's creation of primeval slime of *Genesis* 1: 2, his background the creation of light and his foreground the creation of dry land of *Genesis*

Fig. 2.3 Conrad Martens, 1801–1878 *Coochin 27 Nov. 1851* Watercolour. *Augusta Maria Scott: album of watercolour drawings, ca. 1816–1850*. Bequeathed by D.S. Mitchell, 1907. Mitchell Library, State Library of New South Wales. PXC 323, f. 38. Reproduced with kind permission

1: 9–12, and even the garden of Eden of *Genesis* 2: 6–9. Martens also places what looks like a spoonbill in his swamp, the monstrous Stymphalian bird that Hercules is required to kill as one of his seven labours as related in Greek myth (as discussed in later chapters of the present volume).

In another watercolour of 'Coochin Run' painted on the spot in 1851, Martens does depict Aboriginal people subsumed like the swamp as in his later painting within the pastoralist landscape (and so equating swamp Aboriginals with the swamp), this time with a hut and horses in the background to leave the viewer in no doubt about that this is now a colonized pastoralist landscape (and not pastural land) (Fig. 2.3):

The Queensland Art Gallery of Modern Art (n.d.-b) comments that in this painting 'an Indigenous family group with their weapons and dogs is seen close to a shepherd's hut. Some are wearing European clothing.' Martens poses Aboriginal people in a passive, static tableau within the pastoralist landscape against the backdrop of mountains. They are not shown actively practising any of their traditional (wet)land uses. Von Guérard also either depicted Aboriginal people in the foreground of some of his paintings, such as in his famous painting of Tower Hill near Warrnambool in western Victoria, or not at all, as in his painting, 'Mount William from Mount Dryden,' in which their mobs of kangaroos and wallabies are grazing, but from which Aboriginal graziers and pasturalists are absent, or have been absented by being removed.[1]

Conrad Martens' two paintings can be contrasted with the paintings by Mavis Ngallametta of the Kugu-Uwanh people, Putch clan, depicting Queensland swamps. Her traditional and contemporary swamp paintings are exquisite and intimate artworks that embrace both eras. Rather than swamps with their negative connotations, the wetlands they depict are sometimes described as lagoons:

> Mavis Ngallametta / *Little Swamp on the Way to Obun* 2018 / Natural pigments and charcoal with acrylic binder on linen primed in synthetic polymer paint / 271 x 200cm. Accession No: 2018.393 / Collection: Queensland Art Gallery of Modern Art / © the estate of Mavis Ngallametta c/- Martin Browne Contemporary.
>
> Accessed May 13, 2023: https://collection.qagoma.qld.gov.au/objects/32442

The gallery comments on this painting that:

> A popular site for fishing, Obun is a place where the people of Aurukun, in far north Queensland, go to relax and catch food for their families. The little swamp in the painting is one of the main freshwater lagoons found near the massive saltwater estuary of Archer Bay. (Queensland Art Gallery of Modern Art n.d.-a)

These swamps are part of the region of the Gulf of Carpentaria and its catchments discussed in greater detail later in the present chapter.

Mavis said in an artist statement in 2018 that:

> There are plenty of birds there. When it's dry time, you see wallabies in that area, but in the wet time, you see only birds and lily flowers and lots of

[1] For further discussion of von Guérard's paintings of wetlands in western Victoria, see Giblett (2023, Chap. 3). For further discussion of paintings by other painters of wetlands elsewhere in Australia, see Giblett (2020a).

swamp flowers. There are little swamps everywhere when the wet sets in. (cited by Queensland Art Gallery of Modern Art n.d.-a)

The gallery also comments that this painting is 'one of the most refined of the artist's *pamp* (swamp) works, which developed considerably in style over the last decade of her life.'

A decade before she painted another painting of a swamp:

Mavis Ngallametta / *Pamp (Swamp)* 2009 / Synthetic polymer paint on linen / 116 x 111cm116 x 111 cm / Accession No: 2015.096 / Collection: Queensland Art Gallery of Modern Art / © the estate of Mavis Ngallametta c/- Martin Browne Contemporary.

Accessed May 13, 2023: https://collection.qagoma.qld.gov.au/objects/21952

The gallery comments on this painting that this is 'one of the artist's earliest works.' Mavis said in her artist statement that it depicts:

the big swamp near Aurukun. There are many different coloured waterlilies and plenty of birds at the swamp. People used to go to the swamp and get the roots of the waterlilies for food and medicine. (cited by Queensland Art Gallery of Modern Art n.d.-a)

The gallery comments that:

Many associate swamps with stagnant water, algae and putrid mud. In contrast, Ngallametta's swamps are picturesque lagoons that ring the community of Aurukun in the post-wet season. In these lagoons, crystal clear fresh water abounds, tainted only by the tea-coloured tannins of the melaleuca trees lining the banks. (Queensland Art Gallery of Modern Art n.d.-a)

The gallery acknowledges swamps' negative associations and lack of pleasing qualities in the minds and for the bodies of many gallery-goers, members of mainstream western culture. The gallery then gives a revisionist reading Mavis of Ngallametta's lagoon paintings as positively picturesque, not pejoratively swampy.[2] The good, crystal clear fresh water of the

[2] The picturesque is a problematic category in mainstream western aesthetics, just as much as the sublime and the beautiful. For further discussion of the picturesque, see Giblett (2011, part II). The sublime is discussed in greater detail in a later chapter of the present volume.

lagoons is posed against the bad, stagnant water and putrid mud of swamps. Yet lagoons and swamps are both wetlands, and wetlands can have both sorts of water. They should be appreciated for both and not moralized in a Manichean, fundamentalist and triumphalist conflict of good versus evil. The title of Ngallametta's *Sacred Swamp* of 2008 says it all. Swamps are sacred places, and sacred places are her Country. She said, 'I paint my Country—the places where we camp, fish, collect ochre and clay and the beach hibiscus that I burn to make charcoal. My paintings tell my grandchildren and their grandchildren the stories of those places and what we do there, the animals and birds and fish that live there, the plants that grow there.'

Unlike Martens' painting with the swamp subsumed within the pleasing picturesque prospect of a pastoralist landscape with a sublime mountainous backdrop bathed in beneficent light from which the viewer is detached and masters from a distance, Mavis Ngallametta's paintings take the aerial and close up view of swamps or lagoons and produce two exquisite and intimate artworks that invite the viewer to wade into the wetland and experience it bodily through the senses of sound, smell, touch and sight. Whereas Martens' paintings depict the landscape in the vertical plane as picturesque for the masculinized viewer to master, Mavis Ngallametta's paintings portray (they are not landscapes) wetlands in the horizontal plane as exquisite (to use Rebecca Solnit's apposite term). Solnit applies the exquisite to contemporary artworks of female bodies, but it can equally be applied to contemporary artworks of feminized bodies of water, such as swamps and lagoons (Solnit 2003, 205–218).[3] Mavis Ngallametta's paintings look like embroidery. The surface of her paintings and the surface of the lagoons are embroidered with flowers.

2.2 'A Swamp is a Fascinating Place'

Eighty years after Morrill and Martens, white people still needed convincing that swamps are any good to them, or at least for waterbirds, if an article in the Brisbane *Telegraph* is anything to go by. On Saturday 11, April 1931 under the by-line 'Mopoke,' in the presumably regular column

Edmund Burke and Immanuel Kant contrasted the sublime with the beautiful. For further discussion of both, see Giblett (1996, Chap. 2).

[3] For further discussion of Rebecca Solnit, the patron saint of progressive politics, see Giblett (2021, Chap. 38).

headed 'Nature Notes' and under the title 'Around Queensland Swamps,' the writer begins by acknowledging upfront the negative connotations of:

> the word 'swamp' [that] may bring to mind low-lying and uninviting flats, flooded and boggy and altogether to be avoided. Sour land and waste land it may be, but to me a swamp is a fascinating place—and very often quite a beautiful one. ('Mopoke' 1931, 10)

Like the Queensland Art Gallery Museum of Modern Art in its commentary on Mavis Ngallametta's paintings, 'Mopoke's' opening gambit acknowledges upfront the swamp's lack of conventional aesthetic qualities in the minds and for the bodies of his/her readers, members of mainstream western culture.

'Mopoke' goes on to indicate that she or he is not only a nature lover in general, but also in particular a paludiphile, a lover of swamps:

> One has to be enthusiastic enough to brave mud and slush and mosquitoes usually, but if the nature lover is keen enough to do this, he [*sic*] will never regret his explorations of swamplands.

'Mopoke' invites his/her readers to wade into wetlands and become a paludiphile with him/her.

'Mopoke' goes on to list the waterbirds that nest 'in the low-lying, semi-dry patches within Queensland swamps' including black swans, brolgas and spoonbills, seemingly depicted in Conrad Martens' painting *Coochin Run* wading in the swamp and the monstrous birds that heroic Hercules of Greek and Roman myth has to kill as one of his seven labours (as we will see in later chapters).

'Mopoke' makes special mention of:

> the pretty nankeen night heron [who] arrives there about sundown for its night fishing, just when others are leaving the place: and even throughout the darkest hours of night the life of a swamp is astir.

The nankeen night heron is a nocturnal fisher, a creature of darkness. The life of a swamp is more active at night than during the day. Like Henry David Thoreau, 'the patron saint of swamps,' 'Mopoke' loves nature at night. Thoreau also loved moonlight (Thoreau 1863). 'Mopoke' hears the swamp astir at night, whereas Thoreau aspires to have all his senses still (Thoreau 1982, 382).

'Mopoke' goes on to argue that the swamp is:

> a veritable living museum, truly, ready for him [*sic*] who is willing to spend a patient and somewhat muddy holiday among the fastnesses of a swamp.

'Mopoke' wanders off the topic of swamps onto other ones, presumably in order to fill column inches, before drawing some general conclusions from observing the lives and habits of birds that:

> there is real organization in most cases; mutual aid instead of competition; united effort, and therefore higher development and more chances to survive. In nature things prey upon one another, but animal socialibility [sic] is just as much a law.

'Mopoke' not only acknowledges that competition is a part of the law of nature, but also cooperation is just as much a law of nature.

'Mopoke' is understandably a reader of Charles Darwin on evolution, natural selection and the survival of the fittest. S/he is perhaps an unlikely reader of Prince Peter Kropotkin on mutual aid as a factor in evolution. For Kropotkin (1989, 6) 'mutual aid is as much a law of animal life as mutual struggle, but [...] as a factor of evolution, it most probably has a far greater importance.' Tim Flannery (1994, 15 and 84) concurs that:

> evolution in Australia is not driven solely by nature 'red in tooth and claw'. Here, a more gentle force—that of coadaptation—is important [...] It is cooperation rather than competition which has been selected for in many Australian environments.

Nature is not only 'red in tooth and claw,' as Alfred Lord Tennyson put it, but also green in leaf and branch. For Kropotkin (1989, 30 and 60), 'the war of each against all is not *the* law of nature [...] Life is struggle; and in that struggle the fittest survive,' but the fittest (the survivors) are those who are the most cooperative, not the most competitive. The law of nature is each with all, for each is all. 'Mopoke' completely ignores or overlooks Aboriginal peoples in Queensland who appreciated and used swamps sustainably for tens of thousands of years before colonization (as we have seen and will see again later in the present chapter).

2.3 Judith Wright: Patron Saint of Australian Conservation

Appreciation for iconic environments in Queensland, such as the Great Barrier Reef and Daintree Rainforest, and efforts to conserve them, has meant that Queensland wetlands have been neglected, overlooked or ignored for too long. An exception is Judith Wright (2015–2000) who lived much of her life in Queensland and was also a leader in the campaigns to 'Save the Great Barrier Reef' (Wright 2014) and for Aboriginal land rights (Wright 1991). She is the patron saint of Australian conservation.[4] In an article published in 1982, Wright (1992, 94) critiqued 'the draining of marshland, swamps and lagoons which seem to have bordered most watercourses [in eastern Australia] in early [pre-colonial] times, taking the overflow of heavy rainfall, and releasing it slowly back as the main stream sank.' Also known as billabongs and floodplains, where they have not been drained or filled these water-bodies bordering watercourses perform the vital twin roles of water storage for times of drought and flood mitigation in times of heavy rainfall. Ignoring this role and building houses on floodplains is not a good idea as many towns and cities have found and still find to their peril.

A year later in 1983 Wright (2022, 257) contributed a chapter devoted to the topic of 'Australia—Landscape Ancient and Modern' in which she was more specific about wetlands in Queensland and the Northern Territory. She highlighted 'the marshes and mangrove swamps of the southern coast of the Gulf [of Carpentaria],' their importance for waterbird breeding and 'spectacular for their great beds of waterlilies.' She also critiqued the pastoralist industry with its cattle and buffaloes that 'trample and drain them in the dry season' (Wright 2022, 257). Cattle, buffalo and sheep were the foot soldiers of the colonial invasion. Their hard hooves trampled native vegetation and ruined wetlands, unlike the soft foot pads of native animals. They destroyed the pastoral/ist landscapes beloved of landscape painters, such as Martens, that hid the destruction and aestheticized the land for the visual consumption of the pastoralists—a plain enough case of having your cake and eating it too.

[4] For further discussion of Judith Wright as the patron saint of Australian conservation, her leading role in the campaign to 'Save the Great Barrier Reef' and her activism for Aboriginal land rights, see Giblett (2020c, Chap. 9).

A couple of years later in 1985 Wright critiqued 'this very word landscape' arguing that it is 'a painter's term, implying an outside view, a separation, even a basis of criticism. We cannot set it against the reality of that earth-sky-water-tree-spirithuman complex existing in spacetime, which is the Aboriginal world' (Wright 2022, 272). Unlike 'Mopoke' and Winter, Wright did not ignore or overlook Aboriginal people as she had a strong appreciation for Aboriginal Country, including the Rainbow Serpent that she called or referred to as 'the Great Rainbow Snake.' Wright (1991, 138) pointed out that 'the country is made up of songs and stories linked across tracks and territories, and of the bodies of its makers,' such as this Snake. The earth sings its own songs. Wright heard them and wrote many songs in its song-book in her prose and poetry.

Conservation activism and writing a word for wetlands come together in Wright's prose and poetry. Wright was also a cultural botanist who celebrated wetland plants in poetry, such as *Mazus pumila* in her poem 'Swamp Plant:'

> I have seen those very seldom-seen plants:
> small earth-hugging rosette,
> stem like a thread and downward-turning bell
> of meditative blue. Half-size to a grasshopper,
> what insect is small enough
> to drink from you? *Mazus*,
> *Mazus pumila*, somebody saw and named you.
>
> Only science, then, has noticed you,
> not poetry.
> It's that way round in this country,
> upside-down as ever.
> Living on swamp-edges
> turning your face to the ground, shyer than
> Wordsworth's violet,
> no words but dog-Latin
> have tagged you.
>
> But for your colour—
> such a colour as old sea-goddesses chose
> (Mediterranean goddesses)
> I would not have stooped to look.

But there are no flowers here
Persephone could have gathered;
nor do our people go
down on their knees at swamp-edges
or shorten their range of sight
to your less-than-a-finger's height.

Leaving you there, I take you home with me,
one tiny image
of still untouched unknown tranquillity. (Wright 1976, 40–41; 1994, 367)

Mazus pumila is not native to Australia and is an invasive species, in other words a 'weed.' Wright celebrated the beauty of this swamp plant without being concerned about this distinction, or being aware of the plant's origin. Perhaps for these reasons, Wright did not include this poem in her own selection of her poems for *A Human Pattern* (Wright 1996).

2.4 Queensland Wetlands and Waterbirds

Published in the 1980s like Wright's essays and 50 years after 'Mopoke's' article, J. W. Winter follows 'Mopoke' when he largely ignores or overlooks Aboriginal peoples in Queensland. Winter contributed a chapter on 'The Swamps: A Habitat in Motion' to an edited collection devoted to the natural history of Queensland (Winter 1983). Winter focusses on swamps as waterbird habitat, rather than as homes for Aboriginal people, though he does acknowledge that swamp plants were a food source for Aboriginal people, albeit in a caption to a photo:

> the tubers [of the bulkuru sedge *Eleocharis dulcis* (water chestnut)[5]] are rich in carbohydrates and form the major food supply of both brolgas and magpie geese (and formerly, Aboriginals). Bulkuru is a tall perennial freshwater sedge. It is the most abundant water-plant on the coastal swamps of the Burdekin/Townsville region. (Winter 1983, 180)

He does not describe how bulkuru was cultivated in paludiculture, unlike Morrill. This is Winter's only mention of Aboriginal people.

[5] This edible wetland species is discussed in early ethnobotanical literature using former scientific names; see Roth (1901, 13) and Thozet (1866, 7).

Referring to previous studies of waterbirds in Queensland, Winter highlights the fact that 'the coastal swamps of the Burdekin/Townsville region were identified as amongst the finest wetland habitat for waterbirds in north Queensland' (Winter 1983, 148; see also 150 for more detail). This region of Burdekin River and Townsville is where James Morrill lived with Aboriginal people for 17 years. After being shipwrecked in 1846, Morrill in fact landed at the site of present-day Townsville/Cleveland Bay as confirmed in the indication on a map published as a frontispiece to his book (Morrill 2006, vi). He later described the lives and livelihoods of Aboriginal people living with swamps in this region (as we saw previously in the present chapter).

Winter acknowledges that:

> in Queensland [...] numerous areas of wetland are scattered throughout its length and breadth. Many are ephemeral; few have never been known to dry out. The main concentrations of wetlands are on the coast [...] The inland swamps are vast in wet years [...] These swamps are quite different from the coastal swamps, where regular annual oscillations of wet and dry are important. (Winter 1983, 148; see also 183)

Oscillations of wet and dry are important for inland swamps too, but they are neither regular, nor annual.

Winter concludes with a call for conservation of the coastal swamps of the Burdekin/Townsville region:

> The justification for their continued existence is steeped in the philosophy of conservation and in the growing awareness that regardless of his [*sic*] environment-altering capacities, man is still part of the natural order. (Winter 1983, 186)

Indeed. As Rachel Carson (1994, 86 and 188), the patron saint of American conservation, wrote in *Silent Spring* 20 years before Winter, 'man is part of nature.'[6]

The Queensland Museum in the book *Wetlands of Queensland* concurs recently with the Aboriginal people of the Burdekin/Townsville region that swamps are good and useful, that they have traditional and spiritual significance, and that they provide physical sustenance (without

[6] For further discussion of Rachel Carson as the patron saint of American conservation and *Silent Spring*, see Giblett (2020c, Chap. 8).

acknowledging paludiculture much, or the cultivation of wetlands as such). It also elaborates on Winter's discussion of wetland types and specific wetlands in Queensland, including in the Burdekin/Townsville/Cleveland Bay region and Bowling Green Bay, east of Cleveland Bay (Queensland Museum 2022, 344–349). It also acknowledges and discusses Aboriginal people's sustainable use of wetlands. Aboriginal people in Queensland cultivated the abundant animal and plant life of wetlands for food for many thousands of years in paludiculture. It concurs with Winter that bulkuru is a food source for First Nations' peoples and waterbirds (Queensland Museum 2022, 295).

Wetlands of Queensland, published by the Queensland Museum in 2022, is a lavishly illustrated book and weighty tome of over 400 page that the sets a benchmark for the other state museums in Australia and should be a source of envy for them. It documents and presents extensively the ecological values of wetland types and selected individual wetlands in five regions of Queensland. It also discusses briefly and periodically the role of these wetland types and individual wetlands in the lives of Traditional Owners/First Nations People. Yet there is no discussion of the people who colonized and destroyed wetlands in Queensland. It is as if the only people in Queensland are Aboriginal peoples. Non-aboriginal people and their role in the colonial, agricultural and urban destruction of wetlands in Queensland, and their conservation, are written out of this account.

The only evidence for this history are aerial photographs of urban areas abutting wetlands (Queensland Museum 2022, 218–219, 221) and of 'canal estates,' or 'modified wetlands' (221), and the brief acknowledgement that:

> most of the human population of Southeast Queensland lives on lower floodplains and coastal areas, and consequently these areas are highly modified. Many wetlands in these areas have been cleared, drained or filled. (Queensland Museum 2022, 221)

Or modified into canal estates, in a word, 'canalized.' One wonders when were these wetlands cleared, drained, filled or canalized? Why? By whom? Presumably by 'the human population of Southeast Queensland,' though not all members of this population are equally culpable, certainly not Aboriginal or First Nations people or Traditional Owners. Colonial history is largely written out of *Wetlands of Queensland,* including the present when history is in the making. For instance, the Queensland

Museum does not mention nor discuss the development proposal to modify the Ramsar-listed wetlands of Toondah Harbour and their environs in Southeast Queensland by constructing even more canal estates. Nor does it is discuss the campaign to save these wetlands from destruction (Readfern 2023).

By contrast with the unnamed colonial wetland drainers, fillers or modifiers, early on *Wetlands of Queensland* states that:

> wetland ecosystems are of profound material and cultural importance to First Nations people. Almost all wetland plant and animal species have some form of traditional use—particularly vegetation, crustaceans, fish, reptiles, mammals and waterbirds (especially their eggs)—or are valued for their cultural significance. Wetlands supply medicine and tools for Traditional Owners [...] Wetlands are also story places and centres for cultural activity [...] Aboriginal knowledge of wetland management provides an important basis for natural resource management, which has evolved over several hundred generation of people living on and managing custodial responsibility for country. (Queensland Museum 2022, 18)

In other words, Aboriginal knowledge of wetlands is reduced to its contribution to the neo-colonial and institutional discourse of 'management.' Aboriginal wetland people are dispossessed from their traditional ways of life gaining sustenance from wetlands unless they can demonstrate 'Native Title' to the National Native Title Tribunal and Federal Court by proving traditional ties of hunting, gathering and foraging, and ceremonial uses, from which colonist have largely dispossessed them. Paludiculture does not count as a test of 'Native Title' under the current legislation.

Succeeding chapters of *Wetlands of Queensland* are devoted to discussing the wetlands of five regions of Queensland, beginning with Lake Eyre-Kati Thanda, 'one of the world's greatest desert river basins' (Queensland Museum 2022, 105). These chapters include a brief discussion that reiterates and elaborates the general statement cited previously in relation to Traditional Owners and the wetlands of the region, including in this case their spiritual significance:

> Traditional Owners have social, economic, cultural and spiritual affiliations with and responsibilities for the land and waters of the Lake Eyre basin [...] For the Traditional Owners, caring for country is more than a matter of economic prosperity, it is a scared and ancient traditional responsibility carried forward from mother to daughter, father to son, and includes social,

> environmental, and cultural considerations. Traditional people live by the seasons and think of country as their mother and of water as the sacred life-blood keeping them connected through hunting, fishing and ceremonial practices. (Queensland Museum 2022, 105)

It also keeps them alive through these and other practices such as foraging and gathering. Water is the lifeblood of country and its people that flows above ground in rivers, creeks and wetlands and below ground in aquifers. The Queensland Museum (2022, 71) points out that 'the Lake Eyre Basin overlies the Great Artesian Basin, one of the world's largest aquifer systems.' Wetlands are the surface expression of subsurface water-bodies. Water is the lifeblood of the body of the earth (see Giblett 2007, 2009, Chap. 11).

The neo-colonial and institutional discourse of management in *Wetlands of Queensland* evident in the chapter on Lake Eyre-Kati Thanda persists in the following chapter in discussing the Budjiti people and the wetlands of Currawinya National Park. These 'nationally and internationally renowned wetlands' are part of the Paroo River country, in turn, part of the Murray-Darling Basin (Queensland Museum 2022, 180). These wetlands 'form part of Budjiti cultural landscape' (180). In other words, they also form part of the colonizer's natural landscape. Aboriginal people and colonizers are made to sit either side of the nature/culture divide with land regarded as cultural by the former and as natural by the latter that denies the work of the former and their ancestral beings in creating the land and 'managing' it. As the Queensland Museum goes on to state:

> for thousands of years before the colonization of the Paroo River country, Budjiti peoples sustainably managed their ancestral lands and waters. (Queensland Museum 2022, 180)

Yet more than merely 'managing' ancestral lands and waters for their own sake, or only for their environmental 'sustainability,' these peoples also used sustainably their 'resources' for human sustenance and health:

> of great cultural significance to the Budjiti peoples, the lakes, springs and other wetlands of Currawinya National Park provide an important source of food, medicines and material resources. Budjiti Ancestors built *dikinj* (camps) by the lakes and made nets to catch waterbirds, including *guturu* (swan), *yuli* (wood duck), *mingara* (black duck) and *gultaba* (teal duck). They also collected the bird eggs and fished in the water, catching *gupiri*

> (black bream), *bugili* (crayfish) and *birriri* (turtle). (Queensland Museum 2022, 180)

Budjiti are settlers who built camps and who practised aquaorniculture and aquaculture, and possibly paludiculture too as:

> These places with the cultural landscape—including birthing and burial places, hunting grounds, sowing and harvesting areas, fish traps, campsites and ceremonial grounds—give shape and form to the stories, traditions and memory of Budjiti peoples, and support the continuity of these cultural practices into the future. (Queensland Museum 2022, 180)

Which species of plant are, or were, sown and harvested is not specified. The cultural practices of aquaorniculture and aquaculture, and possibly paludiculture, in the places in which they are performed also give shape and form to the stories, traditions and memory of Budjiti peoples.

The Butchulla people of K'gari (Fraser Island, 'the largest sand island in the world'; Queensland Museum 2022, 190) have similar practices to the Budjiti as evidenced by 'fish traps, midden sites, stone scatters, knapping floors (tool shops), stone tool kits, ceremonial areas, spiritual areas, pathways and scar trees' (Queensland Museum 2022, 258). These relics are evidence not only that the Butchulla people 'occupied K'gari for many thousands of years' and 'for having lived in this area,' but also that they used and 'managed' its 'resources' in their livelihoods. They not only occupied and lived in this place, but also used and thrived in it. They are less a wetland people (given the location and nature of the terrain of a sand island) and more of a 'land and sea people' with 'a strong connection to sea country' (Queensland Museum 2022, 258). These and other Aboriginal peoples elsewhere in Australia regard the sea as part of their country and refer to their country in the sea (Webb 1996, 69) and to their sea country (Giblett 2023, 182–184).

The Wanjuru-Yidinjii peoples of the Russell River catchment practised aquaorniculture and paludiculture in Eubenangee Swamp before the arrival of the colonizers. In this swamp 'they burnt the grasslands to facilitate the growth of fresh shoots in order to attract wallaby, ducks, geese and other animals' that they would 'hunt and trap' (Queensland Museum 2022, 326). The swamp 'provides a rich hunting and fishing ground within the catchment' that:

> revolves around ganyarr [the saltwater or estuarine crocodile] and the wet and dry seasons. Ganyarr permanently reside in Eubenangee but are more active during the wet season when breeding occurs […] Wanjuru people share Eubenangee with ganyarr during the relative safety of the dry season. (Queensland Museum 2022, 326)

Ganyarr is the apex scavenger of the swamp that occasionally takes and consumes with regularity a human that ventures into its territory. Saltwater crocodiles have been typecast as monsters, regarded with horror and associated with the uncanny.[7]

Jaban, the freshwater eel, resides in the upper catchment of Russell River and is 'the keeper of water' as it 'keeps the water clean.' The Queensland Museum goes on to argue that:

> Protecting jaban—the apex predator in the upper reaches of the catchment—means caring for the waterways, which results in healthy aquatic ecosystems for all other species and, in turn, plentiful supply of food for Wanjuru. (Queensland Museum 2022, 326)

Healthy land and water mean healthy people; no healthy people without healthy land and water. Human health includes mental and spiritual health.

The Queensland Museum concludes of this swamp that:

> Today, Eubenangee is an ecologically significant palustrine wetland, and one of the last remaining associations of specific ecological communities (vine forest, grassland, sedge and melaleuca paperbark swamp forest) occurring on nutrient-rich basaltic alluvium. (Queensland Museum 2022, 326)

All good reasons for caring for and conserving Eubenangee Swamp.

In the far north of Queensland, the southern Gulf of Carpentaria is:

> particularly significant for the extent and continuity of its wetlands. These wetlands are important habitat for many waterbird and shorebird species and have cultural significance to Traditional Owners. (Queensland Museum 2022, 412)

[7] For further discussion of monstrous crocodiles, horror and the uncanny, see Giblett (2018, Chap. 4; 2019a, Chap. 2; 2019b, Chap. 3).

For Hayden Tyrrell, the Normanton Ranger Coordinator and a Kukatj man of the Wernadinga coastal site, 'wetlands means life [...] We value wetlands for our culture and our future' (cited by Queensland Museum 2022, epigraph front cover flap and 412).

Prominent among the wetlands of the Southern Gulf for the Kurtijar peoples is the Delta Downs flyway site with 20,000 hectares of wetlands and over 20,000 migratory species of waterbirds from at least 20 species (Queensland Museum 2022, 412). The twice as large Nijinda Durlga (Tarrant) wetland site north of Burketown is part of an Indigenous Protected Area with over 10,000 migratory shorebirds coming and going. It is managed by the Gangalidda and Garawa people whose ancestors collected shell-fish as evidenced by their middens and told by their stories. Desmond Armstrong, a Gangalidda man and the Gangalidda and Garawa Ranger Coordinator relates that 'today our people continue to camp in these areas, as they are excellent hunting and gathering locations and important story places where traditional ecological knowledge is shared' (cited by Queensland Museum 2022, 412).

Those stories about these places and traditional ecological knowledge are shared in *Wetlands of Queensland*. They promote awareness and understanding of the importance of Queensland's wetlands, 'both for the role wetlands play in the surrounding environment, and for our [human?] reliance upon them for our well-being' (Queensland Museum 2022, 419). Wetlands are vital for the well-being (aka mental health) of Aboriginal and non-Aboriginal peoples. As Hayden Tyrrell says, 'wetlands mean life.' No wetlands, no life; no wetlands, death in a word.

2.5 Queensland Wetlands in Literature

The wetlands of Queensland are one of the settings for Alexis Wright's uniquely Aboriginal Australian magic realist, dystopian and climate change novels, *Carpentaria* and *The Swan Book*. In particular, the wet and dry country of the Gulf of Carpentaria is the precise setting for *Carpentaria*. It begins by relating how:

> the ancestral serpent, a creature larger than the storm clouds, came down from the stars, laden with its own creative enormity [...] It came down those billions of years ago, to crawl on its heavy belly, all around the wet clay soils in the Gulf of Carpentaria. Picture the creative serpent scoring down into—scouring down through—the slippery underground of the mudflats [...]

> The serpent travelled over the marine plains, over the salt plains, through the salt dunes, past the mangrove forests and crawled inland [...] where the giant serpent continues to live deep under the ground in a vast network of limestone aquifers. (Wright 2006, 1–2)

This place is otherwise known as 'the depths of the creature of the underworld's belly' (437). The serpent, its belly and the water-bodies of the underworld are contiguous, continuous and consubstantial. They are one body, the water-body of the earth.

As creator of water-bodies and resident of the water-bodies below ground, 'the serpent's covenant permeates everything,' including all aspects of life of the watery place and 'the water people' (Wright 2006, 11) who live there. People and place are bound together in the traditional social and natural contract of the serpent that even permeates the colonial town rubbish dump with 'the haunting spirits residing in the smelly residue, deep down in the gloomy, slime-dripping serpentine caverns of the dump' (16). The subterranean strata of the slimy and smelly town dump are a cruel joke and sick parody of the serpent's residence of limestone aquifers under the ground and the other water-bodies it created above ground.

Similarly a lagoon also becomes a cruel joke and sick parody of creative life as it was traditionally a secret fishing hole for Aboriginal people, but becomes a dumping spot for a dead body. A group of Aboriginal people come to the lagoon, but there was no fishing to be done then as the lagoon was 'shallow and brackish' and 'drying up until the Wet arrived' (146). It is not a swamp of stagnant water and putrid mud for Aboriginal people, but a seasonal wetland for seasonal fishing. This secret fishing hole, 'sacred lagoon' (168) and 'the lagoon of the dancing spirits' (418), or 'stagnant pond' (166) depending on when and how you look at it, is where the local miners have dumped the body of Elias who they have murdered. He is a local non-Aboriginal fisherman and a friend of one of the Aboriginal clans. They have transported his 'tinny' (small fishing boat) up from the coast and have left his body in it. Instead of a place of creative, seasonal life, the lagoon has become a place of destructive, untimely death.

Mining is also destructive, the untimely killer of people and place, ripping ores out of the earth. Mining with its 'heavy machinery churning and gouging into the land' (150) is also a cruel joke and sick parody of the serpent scouring down through the underground. Mining machinery is monstrous and destructive, unlike the serpent who is gigantic and creative. 'Mined ore' is 'the flesh of the earth' (388), not the living body of country.

One of the main Aboriginal characters, Will Phantom, later drags the boat with Elias' body in it up into the hills. He wants to stop the miners from retrieving the body and the boat when they come back to the lagoon to destroy the evidence. He hides the body and the boat in a cave whose:

> walls were covered with ancestral paintings telling stories of human history, made and remade by ochre paints, as the forefathers whispered the charter of the land. Will acknowledged their presence, touched the walls in places to embrace the timelessness of his own being. He felt humbled, honoured to be in the home of birds, animals and clans people of time passed. (182)

This cave is part of 'the country of the water people' (182). Water is not only on and under the ground, but also in the air when it rains and after it stops when Will 'felt cleansed by the refreshing air steaming off the damp earth as it lifted in pencil-thin columns of mist from the sodden bush' (183). This cleansing and health-giving air is not the polluting air of miasma that was thought to cause malaria (literally 'bad air') by the ancient Greeks and their successors (as we will see in the following chapter of the present volume).

The miners cannot find Will and Elias when they search for them on land and from the air. Later Norm Phantom, Will's father, later takes Elias' body out to sea and gives him/it a decent sea burial. The seasonal lagoon plays a vital role in the story. It is like the lagoons in Mavis Ngallametta's paintings, a place of contested meanings, a site of semiotic struggle, as is the mine. Will and a couple of other young Aboriginal men later sabotage and destroy 'the biggest mine of its type in the world' (409). They use fire to ignite fuel and then they:

> watched the fire rage like a monster cut loose from another world. It might even have come from hell. Even the devil himself would have least expected us weak people to have opened the gates of hell. But we watched full of fascination at the fire's life, roaring like a fiery serpent. (410)

The fiery, serpentine monster destroys the monstrous mining machines that consume fuel in their internal combustion engines, or should that be infernal combustion engines? The fiery, serpentine monster of the inferno fights fire with fire. Serpentine monsters have been objects, or more precisely abjects, of fascination and horror for a long time (including dragons breathing fire as we will see in a later chapter of the present volume).[8] The fiery, serpentine monster is a sort of satanic swamp serpent (like John Milton's) born out of the slimy and fiery inferno (like Dante's, as we will see with both in a later chapter of the present volume). The ancestral underworld serpent who created life and lives underground enacts a terrible revenge in conflagration and destruction on the mine, the miners and their monstrous machinery for churning and gouging into the land. The cruel joke is now on them, the sick parodists parodied. The tables have been turned on them and their world turned upside down in carnivalesque subversion using the weapons of the weak, including storytelling, such as in *Carpentaria*.[9]

Like *Carpentaria* with its lagoon in the Gulf of Carpentaria, *The Swan Book* is set largely in a swamp in north Queensland. It begins with a prelude entitled 'Ignis Fatuus,' Latin for 'foolish fire.' This term is used for the will-o'-the-wisp, the light thought to emanate from marshes gases and to be ignited at night. It is also used for the souls of the dead, in this case, the ghost of the old swan woman (Wright 2013, 2). She is the fiery fool who speaks the truth to power, as in Shakespeare's *King Lear*.[10] She is an emanation of 'the backwater swamp' with its 'polluted swamp water' (3) and its:

[8] Abject is Julia Kristeva's (1982) term for that which is between and prior to subject and object, and refuses both.

[9] For carnivalesque subversion as a radical political practice, as distinct from mere literary parodic inversion, see Mikhail Bakhtin (1984). For a weak messianic power, see Walter Benjamin (2003, 390) in his 'Theses on History.' For storytelling, see Benjamin's famous essay 'The Storyteller' published in many collections of his essays, including most recently and handily in 2019 with related essays (Benjamin 2019). For further critique of mining along these lines drawing on the work of other Aboriginal writers, see Giblett (2011, Chap. 9; 2019b, Chap. 5).

[10] 'Speaking truth to power' elaborates the Greco-Roman concept of παρρησία or *parrhésia,* sometimes translated loosely as 'freedom of speech' or 'fearless speech.' It is not to be confused with free hate speech, or hateful free speech, that exercises and reproduces dominant power (patriarchal, political, sexual, financial, physical, cultural, racial, placial, paludal, etc.) over the dominated, marginalised and oppressed. The French genealogist of power-knowledge relations, Michel Foucault (1926–1984), researched and promoted the ancient cultural and political practice of speaking truth to power in the last years of his life in the late 1970s and early 1980s as discussed in his last lectures and books. Foucault is the patron saint of *parrhésia,* the fearless speech of speaking truth to power. For further discussion of speaking truth to power, see Giblett (2021, 91–92).

> dead trees scattered through the swamp, where *dikili* ghost gums old as the hills grew next to a deepwater lake fed by an old spring-spirit relative, until they all slowly died. This happened during the massive sand storms that cursed the place after the arrival of the strangers form the sea. (8)

This cursed place is not only a tree graveyard with ghost gums and the old swan woman's ghost, but also a boat graveyard with 'phantom vessels' (55) and 'ghost ships on the swamp' (175). It is also the homeland of 'the swamp dwellers [who] watched [...] what was happening to their lot' (66). 'Lot' has a lot of meanings in this context, including their lot in life, their mob, their share, their lot in this division of Country, their bad lot, including the fact that 'the old loved weeping spring was now the stagnant water' (66) with 'toxic waste swimming on the surface of the water' (58), together with 'swimming swans. Swamp people regarded this particular sight as something evil, created by devils' (58–59). The swans were engaged 'in an endless activity of sifting waters stuffed with algal blooms, scum on the surface, and slime-covered waterweeds' (70). This swamp has become stagnant water and putrid mud. It is not a healthy home for humans and more-than-humans alike, including swans. It is 'a dead lake' (19), 'the dead swamp' (54), a ghost swamp, 'renamed Swan Lake' (49), the kind of euphemism and marketing moniker beloved of suburban estate developers.

Oblivia, the central character, wonders what the old swan woman meant 'swinging her mantra in some foreign antique language, the one the old sea explorers used when they saw a black swan for the first time, *Rara avis in terris nigroque simillima cygno*' (Wright 2013, 80). This mantra was devised a long time before European sea explorers saw a black swan for the first time. During the second century of the common era Juvenal (1992) in writing his *Satires* (VI, 161–166) used the black swan as a figure for an impossible woman in answer to his hypothetical question:

> 'From all the crowds of women, can you not find one who is decent?'
> Suppose she is beautiful, graceful, wealthy, fertile, and also
> has ancient ancestors dotting her hallway; suppose she is purer
> than any Sabine with streaming hair who stopped a war—
> a rare bird, as strange to the earth as a black swan;
> ['*rara avis in terris nigroque simillima cygno*']
> who could endure a wife who was such a paragon?

Such a woman, in other words, does not exist, because black swans do not exist, and if she did exist, no man could live with her. It took another millennium and more for a European sea explorer to see a black swan in the flesh, and not just in a figure.

The impossible old swan woman gives Oblivia 'the facts' that the Dutchman Willem de Vlamingh and his crew were the first white men who saw a black swan in 1697 off the coast of Western Australia (Wright 2013, 81, 181–182, 219). Yet this sighting was not the first time a European had seen a black swan. An earlier, documented sighting by the Dutch took place in 1636. Hubert Whittell (1954, 5) in the section called 'First Report of a Black Swan' in his book on the literature of Australian birds relates how Antonie Caen, the skipper of a Dutch ship, recorded in his log for 5 July 1636 'having seen [...] two stately black birds as large as swans.' Whittell concludes that, 'here, almost certainly, is the first record of the Australian Black Swan.' Peter Young (2008, 27; see also 7 and 183) in his *Swan* book in the Reaktion Books animal series does not seem to be aware at all of Caen's sighting and record in 1636 when he asserts in his book about the swan that black swans were 'first reported to the western world by the Dutch explorer Willem de Vlamingh in 1697.' Wright in her *Swan Book* seems to be reliant on Young's *Swan* book for this misinformation. She does not cite a source for it, or him, in her 'Note on Sources' which are limited to 'quotations embedded in the text' (Wright 2013, 336–337). By the titles Whittell (1954, 5 and 6) gives to two sections of his 'A History of Australian Ornithology' he is emphatic that Caen recorded the 'first report of a black swan' in 1636 and equally emphatic that de Vlamingh recorded the 'first capture of black swans' in 1697.[11]

Black swans play a bigger role in *The Swan Book* than as a contentious topic or as a matter of idle curiosity in a natural and cultural history lesson. The local federal politician, Warren Finch, 'the postmodernist, deconstructionist champion' (119), returns to the homeland of the black swans and the swamp people from where and whom he came. Seeing 'the junk in the swamp was not a glorious sight' (117) and smelling 'the fetid smell of the swamp [...] was [...] off-putting' (129). He is an anti-paludalist. He cannot resist punning about 'the swamp of old government policies' (120). They have been superseded by 'the abstract logic of public official language' (295) antithetical to the concrete sliminess of 'the dismal

[11] For further discussion of the natural and cultural history of the black swan, see Wright (2013, 41–42 and 219) and Giblett (2013, Chap. 13).

swamp' (147), its people, their stories in their traditional languages and black swans.

Finch claims that Oblivia has been promised to him and he invokes northern European swan mythology. He calls her:

> the swan maiden [...] It amused him to cast himself into the story found across the northern hemisphere of the hunter who captures a mythical maiden in a marsh. (154)

He and his minders/advisers abduct her and take her to the city in the south where he lives, but not without her reciting 'the many swan maiden fantasies that have vanquished men who hunt swans' (156). Oblivia fights fantasy with fantasy.

One of the most famous of these fantasies is in the Finnish epic poem *Kalevala* in which the heroic Lemminkäïnen wants to marry a beautiful maiden. Her mother gives him the conditional task 'to shoot the beautiful bird,/Shoot the swan of Tuonela/Swimming on the death-dark river' (Lönnrot 2021, 156). On his quest to kill the swan he is killed by a water dragon, but later resurrected by his mother in what seems like a revisionist retake on the earlier story.

Oblivia cannot fight Finch's political power and military might when he orders 'the total evacuation of Swan Lake' and 'the swamp dredged' (160). Finch, the anti-paludalist, commits aquaterracide, the killing of a wetland, an ecological crime. After he is assassinated, perhaps in revenge, Oblivia returns from 'the ghost city' (258) in the south, 'a city of ghost swamps' (Giblett 2020b, part I), to her homeland in the north. The ghost city in the south in *The Swan Book* contains 'the marshy lake in the botanical gardens' (Wright 2013, 276). The Royal Botanic Gardens in Melbourne were founded on a site that included a lagoon, which was landscaped into a pond and recently installed with floating wetlands.[12] In *The Swan Book* the landscaped pond returns to its pre-colonial origins as a marshy lake and lagoon, a case of the return of the repressed, and oppressed.

Oblivia returns to her homeland that was a swamp of ghost gums and ghost ships and is now 'the ravaged landscape that had once been a swamp' (330). Homeland has become a 'wasteland' (240) with 'ghost swans' (329). Oblivia, Finch's 'swan maiden,' has become 'a light moving over

[12] For further discussion of the history of the Royal Botanic Gardens in Melbourne, see Giblett (2020b, Chap. 7).

the marshes in the middle of the night like a will-o'-the-wisp' (334). *The Swan Book* returns to where it began with Oblivia as the new swan woman emanating from the ghost swamp, like the old swan woman, and speaking truth to power. *The Swan Book* also ends on the hopeful note that 'the Rainbow Serpent' might restore the swamp and the 'swans might come back' (334). *Carpentaria* begins with invoking the Rainbow Serpent; *The Swan Book* ends with the same invocation. The Rainbow Serpent lives![13]

From James Morrill's sojourn for 17 years with Aboriginals in Queensland swamps, through Conrad Martens' picturesque pastoral landscapes, Mavis Ngallametta's exquisite lagoons, 'Mopoke's' love and plea for Queensland swamps, J. W. Winter's scientific documentation of them as waterbird habitat, the Queensland Museum's lavish presentation to Judith Wright's pro-paludal writings and Alexis Wright's reworked magic realism of a lagoon and a swamp, the wetlands of Queensland have many facets and features that have been foregrounded or neglected over time. These include low-lying and uninviting flats, flooded and boggy, as well as stagnant water, putrid mud, homeland, sacred lagoon and old loved weeping spring. Wetlands in Queensland, in many other places in Australia and around the world, have been cultivated, invaded, colonized, enclosed, aestheticized, trampled, dredged, drained, filled, canalized, embanked, impounded, studied, loved, revered, portrayed exquisitely and illustrated lavishly. Perhaps no other landform and water-body has had to endure such an array of actions, or been subjected to such a range of ways of being treated. Within this contested space and running through this swampy journey is a call for the conservation of wetlands, as well as the critique of anti-paludal cultures and the promotion of pro-paludal cultures, a call taken up in succeeding chapters of the present volume.

References

Bakhtin, M.M. 1984. *Rabelais and His World*. Trans. H. Iswolsky. Bloomington, IN: Indiana University Press.

Benjamin, Walter. 2003. *Selected Writings: Volume 4, 1938–1940*, ed. Howard Eiland and Michael Jennings. Trans. E. Jephcott et al. Cambridge, MA: Belknap Press of Harvard University Press.

———. 2019. *The Storyteller Essays*. Trans. T. Lewis. New York: New York Review Books.

[13] For further discussion of the Rainbow Serpent, see Giblett (2020d).

Carson, Rachel. 1994. *Silent Spring*. Boston, MA: Houghton Mifflin. First published 1962.
Flannery, Tim. 1994. *The Future Eaters: An Ecological History of the Australasian Lands and People*. Chatswood, NSW: Reed.
Giblett, Rod. 1996. *Postmodern Wetlands: Culture, History, Ecology*. Edinburgh: Edinburgh University Press.
———. 2007. Black and White Water. In *Fresh Water: New Perspectives on Water in Australia*, ed. Emily Potter, Alison Mackinnon, Stephen McKenzie, and Jennifer McKay, 31–43. Carlton: Melbourne University Press.
———. 2009. *Landscapes of Culture and Nature*. Basingstoke: Palgrave Macmillan.
———. 2011. *People and Places of Nature and Culture*. Bristol: Intellect Books.
———. 2013. *Black Swan Lake: Life of a Wetland*. Bristol: Intellect Books.
———. 2018. *Environmental Humanities and Theologies: Ecoculture, Literature and the Bible*. London: Routledge.
———. 2019a. *Environmental Humanities and the Uncanny: Ecoculture, Literature and Religion*. London: Routledge.
———. 2019b. *Psychoanalytic Ecology: The Talking Cure for Environmental Illness and Health*. London: Routledge.
———. 2020a. Artist and Swamp: Wetlands in Australian Painting and Photography. In *Australian Wetland Cultures: Swamps and the Environmental Crisis*, ed. John Charles Ryan and Li Chen, 51–70. Lanham, MD: Lexington Books.
———. 2020b. *Modern Melbourne: City and Site of Nature and Culture*. Bristol: Intellect Books.
———. 2020c. *New Lives of the Saints: Twelve Environmental Apostles*. Lanham, MD: Hamilton Books.
———. 2020d. Rainbow Serpent Anthropology, or Rainbow Spirit Theology, or Swamp Serpent Sacrality and Marsh Monster Maternity? In *Australian Wetland Cultures: Swamps and the Environmental Crisis*, ed. John Charles Ryan and Li Chen, 33–50. Lanham, MD: Lexington Books.
———. 2021. *Black Swan Song: Life and Work of a Wetland Writer*. Lanham, MD: Hamilton Books.
———. 2023. *Middlemarsh: The Hopkins River, Kindred Wetlands and Remarkable People*. London: Transnational Press.
Juvenal. 1992. *Satires*. Trans. N. Rudd. Oxford: Oxford University Press.
Kristeva, Julia. 1982. *Powers of Horror: An Essay on Abjection*. Trans. L. Roudiez. New York: Columbia University Press. First published in French in 1980.
Kropotkin, Peter. 1989. *Mutual Aid: A Factor in Evolution*. Montreal: Black Rose Books. First published 1902.
Lönnrot, Elias. 2021. *Kalevala: The Epic of the Finnish People*. Trans. E. Friberg. London: Penguin. First published in Finnish in 1849.

'Mopoke'. 1931. Nature Notes: Around Queensland Swamps. *Telegraph* (Brisbane), Saturday April, 10, 11.

Morrill, James. 2006. *17 Years Wandering Among the Aboriginals.* Australian Aboriginal Culture Series No.1. Virginia, Northern Territory: David M. Welch. First published 1864.

Queensland Art Gallery of Modern Art. n.d.-a. Learning: Mavis Ngallametta: Pamp (Swamp), Little Swamp on the Way to Obun. https://learning.qagoma.qld.gov.au/artworks/pamp-swamp/; https://learning.qagoma.qld.gov.au/artworks/little-swamp-on-the-way-to-obun/. Accessed 13 May 2023.

———. n.d.-b. Northern Journey: Conrad Martens in Queensland. http://www.visualarts.qld.gov.au/content/martens_standard.asp?name=Martens_Works. Accessed 9 May 2023.

Queensland Museum, in partnership with the Queensland Department of Environment and Science. 2022. *Wetlands of Queensland: A Queensland Museum Discovery Guide.* Brisbane: Queensland Museum.

Readfern, Graham. 2023. Toondah Harbour: Should a Wetland Home to Endangered Birds become $1.3bn Worth of Shops, High-Rises and a Marina? *The Guardian*, May 27. https://www.theguardian.com/australia-news/2023/may/27/should-a-wetland-home-to-endangered-birds-become-13bn-worth-of-shops-high-rises-and-a-marina. Accessed 27 May 2023.

Roth, Walter. 1901. *Food: Its Search, Capture, and Preparation.* North Queensland Ethnography; Bulletin no.3. Brisbane: Government Printer.

Solnit, Rebecca. 2003. *As Eve Said to the Serpent: On Landscape, Gender, and Art.* Athens, GA: University of Georgia Press.

Thoreau, Henry David. 1863. Night and Moonlight. *Atlantic Monthly* 12: 579–583.

———. 1982. *The Portable Thoreau, edited by Carl Bode.* New York: Viking Penguin.

Thozet, A. 1866. *Notes on Some of the Roots, Tubers, Bulbs, and Fruits, Used as Vegetable Food by the Aboriginals of Northern Queensland, Australia.* Rockhampton: W. H. Buzacott, 'Bulletin' Office.

Webb, Hugh. 1996. Aboriginal Country: Not a Construction, A Way of Being. In *Western Australian Wetlands. The Kimberley and South-West*, ed. Rod Giblett and Hugh Webb, 61–76. Perth, Western Australia: Black Swan Press/Wetlands Conservation Society.

Whittell, Hubert. 1954. *The Literature of Australian Birds: A History and Bibliography of Australian Ornithology.* Perth: Paterson Brokensha.

Winter, J.W. 1983. The Swamps: A Habitat in Motion. In *Exploration North: A Natural History of Queensland*, ed. H.J. Lavery, 147–186. South Yarra: Currey O'Neil.

Wright, Alexis. 2006. *Carpentaria*. Artarmon, N.S.W: Giramondo.
———. 2013. *The Swan Book*. Artarmon, N.S.W: Giramondo.
Wright, Judith. 1976. *Fourth Quarter*. Sydney: Angus & Robertson.
———. 1991. *Born of the Conquerors: Selected Essays*. Canberra: Aboriginal Studies Press.
———. 1992. *Going on Talking*. Springwood: Butterfly Books.
———. 1994. *Collected Poems 1942–1985*. Pymble: Angus & Robertson.
———. 1996. *A Human Pattern: Selected Poems*. Watsons Bay: ETT Imprint.
———. 2014. *The Coral Battleground*. 3rd ed. North Melbourne: Spinifex. First published 1977.
———. 2022. *Selected Writings*, ed. Georgina Arnott. Collingwood: La Trobe University Press, in association with Black Inc.
Young, Peter. 2008. *Swan*. London: Reaktion Books.

CHAPTER 3

Marsh Arabs in Iraq

Mesopotamia between the Tigris and Euphrates Rivers is a cradle of western, urban civilizations and cultures. It is, or was, also a wetland as Steve Lonergan and Jassim Al-Asadi (2024, 1) relate in their recent study:

> The Mesopotamian Marshlands were once among the largest wetlands in the world, covering an area of more than 10,500 square kilometers (km^2), roughly the size of Lebanon and larger than twenty-seven countries in the world. During times of extreme floods, the wetlands could extend to 20,000 km^2. They supported a diverse range of flora and fauna and housed a human population estimated between 500,000 and 750,000 by the mid-twentieth century.

The Mesopotamian Marshlands have been the home to the Marsh Arabs for thousands of years and a womb for the creation of their civilization and cradle for its development. They were listed as a UNESCO World Heritage Site in 2016 after a long and tortuous process (Lonergan and Al-Asadi 2024, 157–161).

3.1 The Fertile Crescent and the Garden of Eden

One of the earliest, if not the earliest western outsider to document the life of the Marsh Arabs was John Henry Haynes in 1888 who did so in photography. These photographs are now housed in the University of

R. Giblett, *Wetland Cultures*, Literatures, Cultures, and the Environment, https://doi.org/10.1007/978-3-031-57365-1_3

Pennsylvania Museum of Archaeology and Anthropology in Pittsburgh. A selection of these photographs is included as the final chapter of Edward Ochsenschlager's study of the Marsh Arabs (Ochsenschlager 2004, 251–269). The marsh itself is not photographed as such but relegated to the background as the backdrop for humans, their activities, artefacts and structures. This set the standard for photography for Wilfred Thesiger and Gavin Maxwell to follow in the 1950s.

Following Haynes, in the 1920s the British archaeologist Leonard Woolley conducted research and excavations in southern Iraq that highlighted the fertility of the Marshes (Woolley 1965). Mesopotamia is also known as 'the fertile crescent.' Lonergan and Al-Asadi (2024, 1) relate in their recent study that:

> The region [of the Mesopotamian Marshlands] is part of the Fertile Crescent, a semi-circular cultural and ecological land bridge that ranges from Egypt to Syria to the Gulf. The southernmost section of the Fertile Crescent is also known as the cradle of civilization, where agriculture flourished, and modern culture began. A place where 6,000 years ago, writing, mathematics, metallurgy, and hydraulic engineering were invented, and city-states formed. The most important natural resource that fueled the growth and development of southern Mesopotamia was not oil, but water.

The Fertile Crescent is indeed known by many western scholars as *the* cradle of civilization. Yet civilization was also cradled in other wetlands, such as the Fertile Serpent of western Victoria in Australia where paludiculture and aquatic engineering were invented too (as we saw in the introduction to the present volume). The Fertile Crescent is more precisely *a* cradle of civilization and the cradle of *western* civilization.

Following Woolley, in the 1950s the English explorer and writer Wilfred Thesiger visited Iraq on several occasions where he lived and travelled among the Marsh Arabs and their neighbours, wrote a classic amateur ethnography, *The Marsh Arabs,* and published it with his own photographs in 1964 (Thesiger 2007). Gavin Maxwell travelled to Iraq with Thesiger in 1956 and wrote *People of the Reeds* and published it with his own photographs in 1957, seven years before Thesiger published *The Marsh Arabs.* Thesiger wrote and Maxwell illustrated an article for *National Geographic* (Thesiger 1958). Like Haynes, they did not photograph the marsh itself, but relegated it to the background as the backdrop for humans, their activities, artefacts and structures, with one exception for Thesiger with one out of 110 photographs (number 59) of 'Spring in the Marshes.'

Thesiger's collection of photographs is now housed in the Pitt Rivers Museum in Oxford.

Gavin Young started visiting the Marshes in the 1950s and published his account, *Return to the Marshes: Life with the Marsh Arabs of Iraq* in 1977 with photographs by Nik Wheeler that show the Marshes as active participants in the lives of the Marsh Arabs. Following in Thesiger's and Maxwell's footsteps, these now famous photographs taken for *National Geographic* magazine were first published there to illustrate a story by Young about the Marsh Arabs (Young 1976). Wheeler's photographs were not republished in the 1989 edition of the book by Hutchinson, nor in the 2009 edition republished by Faber, but are easily accessible and viewable through an online search engine. Mootaz Sami's recent monochrome photographs in *The Ghosts of Iraq's Marshes* portray the play of light and dark in the intertwined lives of people and place (Lonergan and Al-Asadi 2024, Figs. 4, 5, 9, 12, 14, 20, 22, 23, 33, 34).

From 1968 to 1990 Ochsenschlager conducted professional academic 'ethno-archaeological fieldwork' in the Marshes of southern Iraq, published articles with his findings in the 1990s, and collected, wrote them up and published them in *Iraq's Marsh Arabs in the Garden of Eden* in 2004 with a selection of Haynes' photographs (Ochsenschlager 2004). Jennifer Pournelle argued in her 2003 PhD dissertation at the University of California in San Diego for the central role of the Iraqi marshlands in the formation of Mesopotamian urban civilization (Pournelle 2003, xxxii–xxxiii). Without the former, the latter would not have been possible. The Iraqi marshlands were the foundation for Mesopotamian cultures and cities.

Ochsenschlager (2004, vii) cautiously followed suit on Pournelle on this point and noted that the Marshes were the site for 'perhaps the world's oldest cities' (without explicitly referencing Pournelle on this point). Ochsenschlager's qualified 'perhaps' indicates that he was not prepared to go down Pournelle's path too far and rock the boat of academic orthodoxy that the Mesopotamian cities were founded on dry land between two great rivers. Pournelle blows this boat out of the water, as this is the main point of her dissertation signalled in her title, *Marshland of Cities: Deltaic Landscapes and the Evolution of Early Mesopotamian Civilization*. Monumental cities made of hard materials can only be built on dry land, but in many cases wetlands were dredged, drained, filled, embanked and/or canalized in the first place to produce dry land (Giblett 2016). Coupled with 'the world's oldest cities,' Ochsenschlager's qualified 'perhaps' also keeps the door open that there are older cities elsewhere in the world. For instance and in fact, the city of Çatalhöyük in Türkiye is older.

Recently James Scott showcased Pournelle's work in two sections of his book *Against the Grain* that I call 'For the Wetlands,' or 'Pro-Paludal' (Scott 2017, 47–57, 127–128). During the seventh and sixth millennia BCE, the rich alluvium of Mesopotamia between the two rivers of the Tigris and Euphrates was what Scott (2017, 47, 127) calls 'a [...] wetland paradise' where marshlanders practised 'an exuberant diversity of livelihoods.' He argues that 'the earliest large fixed settlements sprang up in wetlands' (Scott 2017, 47) in the 'Fertile Crescent' of Mesopotamia. Drawing on Pournelle's work on Mesopotamian marshlands, Scott goes on to relate that these settlements 'relied overwhelmingly on wetland resources [...] for their subsistence' (Scott 2017, 47). Or more precisely, they relied overwhelmingly on wetland resources for their sustenance as these wetlands were traditionally a rich source of animal and plant foods as Scott (47–57, 127–128) goes on to discuss. As the Marsh Arabs regulated the flows of water and cultivated water plants in wetlands, they were not only hunters, gatherers and foragers (as Scott calls the Mesopotamian marshlanders), but also paludiculturalists.

In 2002 and 2003, Jon Lee Anderson visited southern Iraq following in the footsteps of Thesiger and wrote a contemporary, updated account of the Marshes as the introduction to Thesiger's *The Marsh Arabs* for Penguin Classics (Anderson 2007). The Mesopotamian marshlands during Thesiger's time there were, as Anderson (2007, 3) puts in his introduction, 'a 6,000 square-mile [15,500 km^2] area of wetlands created by the Tigris and Euphrates rivers.' Yet as with all wetlands, their surface area fluctuates seasonally and annually as the result of rainfall that increases surface and/or subsurface flows of water to form the wetland (as with Lake Eyre in South Australia as we saw in the introduction to the present volume). Ochsenschlager (2004, 1) concurs in this case and indicates that the Mesopotamian wetlands 'consisted of a series of interconnected permanent marshes and lakes covering some 8,800km^2 [4,400 square miles] in the dry season and expanding to 20,000km^2 [7,800 square miles ...] in the spring inundation.' They are thus smaller and bigger than 6000 square-mile or 15,500 km^2 depending on the season and year in which the area of wet land is measured.

Anderson goes on to point out that many historians believed the marshlands to be the site of the biblical Garden of Eden and Great Flood, and that they had been 'the cradle for human civilization 5000 years before'

Thesiger's visited them in the 1950s. More precisely, the Mesopotamian marshlands are now believed to be a cradle for western, urban civilization as the city of Çatalhöyük in Türkiye is older than the cities of Mesopotamia. Moreover, Australian Aboriginal civilizations began at least 50,000 years before Thesiger visited the marshlands of southern Iraq.

Woolley (1965, 3–4) described how the formation of 'a great lagoon' and 'a waste of water' at the confluence of the Tigris and Euphrates Rivers formed 'a vast delta of clay and sand and mud, diversified by marshes and reed beds [... I]ts soil [...] was rich as could be found anywhere on earth' and scarcely needing labour to produce food. Hence, the area was known as 'the fertile crescent.' He concluded that 'the description in *Genesis* of the creation of the earth as man's [*sic*] home agrees admirably with the process of the formation of the Mesopotamian delta' made up of silt, mud, marsh, lagoon and reed beds. Woolley then quotes *Genesis* 1: 9 and 11, but does not refer to the creation of the Garden of Eden in *Genesis* 2: 8–10 that mentions a river. Ochsenschlager takes up the themes of 'the Garden of Eden' and mud (perhaps an unlikely combination) in his study entitled *Iraq's Marsh Arabs in the Garden of Eden* (Ochsenschlager 2004).

Twenty years later Steve Lonergan and Jassim Al-Asadi bring the story right up to date, including the recent work of rehabilitating the Marshes (Lonergan and Al-Asadi 2024). Jassim Al-Asadi is an environmental activist, head of the environmental group 'Nature Iraq' and one of the foremost experts on the Marshes. It can be a dangerous business in Iraq combining environmental activism and wetland expertise as on February 1, 2023 he was kidnapped as he drove to Baghdad where he was tortured over a two week period. His many contacts in the West, and his public advocacy for the Marshes, made him a target. On February 17, he was released following pressure from his family, friends and outside contacts. Working in the wetlands of Iraq is a lot more risky for Al Asadi than for wetland conservationists living in the West. Steve Lonergan relates these recent events in an 'Afterword' to *The Ghosts of Iraq's Marshes.* He is a Canadian geographer associated with the University of Victoria on Vancouver Island in Canada. The present chapter discusses all of the work mentioned on the Marshes of southern Iraq and the Marsh Arabs with their ancient and traditional wetland culture, including their paludiculture.

3.2 Marsh Arabs in the 1950s

In 1956 Gavin Maxwell travelled to Iraq with Thesiger because he wanted to write a book about the Marsh Arabs and he was fascinated with otters. He travelled through the Marshes with Thesiger in his canoe, or *tarada* (Thesiger 2007, 163). For Maxwell (1966, 34) though, 'canoe is a mean word for so a beautiful a craft' as a *tarada* is 36 feet long and 3 ½ feet wide at its widest point with a 5 feet high prow that enables the Marsh Arabs to force the *tarada* through reed beds where a conventionally shaped boat would get stuck. In addition, their shallow draught of a few inches gives them free passage over water no more than ankle deep (Maxwell 1966, 34). Both men photographed *taradas* and reproduced these in their books.

When it comes to water in southern Iraq, Maxwell privileges on the one hand, 'the two great rivers, the Tigris [with its 'broad and splendid waters' (16)] and the Euphrates' (23), 'the great rivers' (24), 'great waterways' (35) and 'the great reeds of the permanent marsh' (34) and disparages on the other hand 'a watery waste of marsh and lagoon' (10) where the Marsh Arabs 'have for many centuries been proficient in extracting their livelihood from a waste land of water and of reeds' (24). The rivers, the waterways, the waste land and the marsh are all effectively one great wetland (as Pournelle argued and as we will see later in the present chapter).

Maxwell (87) later privileges 'the flowing water' of the Tigris and Euphrates 'as life-blood' coursing through 'the great arteries and veins' of the two rivers. He does not mention that marshes are the kidneys that filter water and the womb and placenta that create life. Lonergan and Al-Asadi (2024, 57) precisely relate that 'the Tigris and Euphrates Rivers are the lifeblood of the Iraqi Marshes.' What Maxwell (27, 46) calls 'the wilderness of reeds and water' and 'this watery waste' sustain the livelihood of the Marsh Arabs. Maxwell describes in detail spearing fish (80–82, 118), herding water buffalo for dung and milk (67–72), and building houses out of reed arches and woven reed mats (29). 'The great *mudhifs*' for Gavin Young (2009, 168) represent 'a uniquely magnificent form of the world's architecture.' Maxwell concludes that 'the reed crop is the marshmen's livelihood' (100). Yet rather than extracting their livelihood from the recalcitrant marsh, the abundant marsh supplies and sustains their livelihood. No marsh, no reeds; no waste land and wilderness, no livelihood; no wetland, no culture; no marsh, no Marsh Arabs. Despite their proficiency in gaining their livelihood, Maxwell (1966, 39) regarded the Marsh Arabs as primitive and contrasted them with civilization. Yet these practices were the basis for their civilization in the sense of making improvements and they demonstrated ownership of the land (wet and dry) as they mixed their labour with nature.

Maxwell (1966, 90–91) appreciated the natural features of the Marshes, such as sunsets. He writes that nowhere had he:

> seen sunsets at once so strange and beautiful as those that flared nightly over the plains of the Tigris and Euphrates [...] Against the still pallid water move the high prows of the canoes [*tarada*], their thin scimitar curves cutting into the embers of the horizon sky. (Maxwell 1966, 90–91)

Maxwell also appreciates 'the strange sounds and sights of the Marshes.' Of them all, the one he thinks he shall:

> remember longest [is] the tumultuous voices of the frogs, those million voices that could turn the great marsh desert into a cauldron of sound seeming more limitless even than the falling horizons themselves. (64)

The Marshes for Maxwell are a cacophonous soundscape and a watery wastelandscape.

Maxwell (38) also appreciates waterbirds, such as pelicans that 'would sail majestically by, riding the wind on stiff outstretched wings, rigid and bulky in body as seaplanes.' They were also 'so oafish and ungraceful on land or water, but so irresistibly glorious in their flight' (73). Pelicans for Maxwell are both ugly and beautiful, a bit like the Marshes themselves. Along similar lines to Maxwell on both pelicans and sunsets, Young (2009, 137) wrote that pelicans, 'ungainly they may be, [...] can be beautiful standing in the shallows to feed at evening' in a flock that 'slowly turns to flamingo-pink as the sun sinks.' They can even be majestic, such as when, 'high up, slowly wheeling and wheeling on outstretched wings, unmoving wings, these sometimes ludicrous-looking birds give to the sweep of water, reeds and towering sky an extra touch of majesty.' Maxwell went on to describe that outside the Marshes 'the semi-inundated rice fields [...] thronged with a multitude of birds: storks, herons, egrets and great flocks of wading birds' (Maxwell 1966, 104). Rice fields, as Emily O'Gorman (2021, 16, 118) says, are 'a kind of wetland,' and all kinds of wetlands are waterbird habitat.

Like Maxwell, but more so, Thesiger was sensitive to the special environment of the Marshes and to the unique features of the life of the Marsh

Arabs. Perhaps that is why he was moved to write his book after Maxwell had written his. Thesiger wrote not only an amateur ethnography about exotic places and peoples that appealed to English readers eager for such heroic ripping yarns, but also a book of environmental writing with an appreciation for the place, its people, plants and animals. *The Marsh Arabs* is usually pigeonholed as a book of travel writing that may have promoted tourism to Iraq when it was possible between intermittent wars and conflicts. Thesiger (2007, 13) himself rejected this classification of his book as he travelled on short day-trips by canoe in 'the restricted area' of the Marshes. Travel writers usually write about going on long journeys from their homeland to and/or in far-flung places.

Thesiger wrote *The Marsh Arabs* after he left Iraq in 1958 and it was first published in 1964. It is in part a memoir of his first visit to the Marshes many years before, in 1951, as 'the memories [...] have never left me' of:

> firelight on a half-upturned face, the crying of geese, duck[s] flighting in to feed, a boy's voice singing somewhere in the dark, canoes moving in a procession down a waterway, the setting sun seen crimson through the smoke of burning reedbeds, narrow waterways that wound still deeper into the Marshes. A naked man in a canoe with a trident in his hand, reed houses built upon the water, black, dripping buffaloes that looked as if they had calved from the swamp with the first dry land. Stars reflected in dark water, the croaking of frogs, canoes coming home at evening, peace and continuity, the stillness of a world that never knew an engine. (Thesiger 2007, 23)

Thesiger paints a romantic and nostalgic picture of the people and the place. He was a primitivist who, in his own words, 'loathed' 'drab modernity,' sought solace in the bygone and the traditional, or 'continuity with the past,' and envied the Marsh Arabs 'a contentment rare in the world today' (Thesiger 2007, 59–60). Yet he also evokes the sights and sounds (though not the smells) of the Marshes in an exemplary piece of multisensorial environmental writing that embraces humans and more-than-humans living in the same place with its unique features. He expresses evocatively that 'I enjoyed being there' (13) in 'a world complete in itself' (20).

Thesiger emphasizes the light, life, spaces and sounds of the Marshes, such as when he stood:

> watching the sun go down behind reedbeds that stretched to the world's end [...] From all around, as if the Marshes breathed, came the massed voices of frogs, an all-pervading pulse of sound, so sustained that the mind ceased to take note of it. More than any other, even than the crying of geese in winter, this was the sound of the Marshes. A dog barked; a buffalo grunted with a noise surprisingly like a camel's; a man called out a long, and to me unintelligible message; a pause, and someone answered [...] A boy, late back from the reedbeds, paddled down a waterway, a path of shining gold leading from the setting sun. He sang softly as he came towards, the notes lingering in the air. (Thesiger 2007, 57)

For Thesiger, the Marshes aren't just a landscape, or even a waterscape or wetlandscape (and certainly not a watery wastelandscape as they were for Maxwell), but also a soundscape and a timescape of the past and present stretching into the future into both of which Thesiger is immersed. Wetlands generally don't lend themselves to appreciation in the conventions of landscape aesthetics addressing the eye, though sunsets in them have their special charm (as Thesiger, Young and Maxwell indicate, and sunrises as in Colin Thiele's *Storm Boy* as we saw in the introduction to the present volume). Wetlands can be appreciated as a living soundscape to the ear attuned to hear it and the voices of its lives and its life as a whole being (ditto Maxwell and Thiele in both *Storm Boy* and *Swan Song*).

Thesiger gives a first-hand, observational and detailed account of the lives of the Arabs in and on the margins of the Marshes, including cultivating rice, managing water levels, enduring flood and drought, herding buffalos, hunting wild boar, constructing elaborate buildings from reeds and sturdy canoes from wood, as well as standing balanced in the front of their canoes and spearing fish. Thesiger says that the Marsh Arabs demonstrated 'a mastery of skills, however simple, I could never hope to attain' (Thesiger 2007, 59–60). Their way of life is a complete paludiculture of cultivating a wetland, which seemed to pass Maxwell by.

Thesiger is also attentive to the waterbirds, including waders, that live in the Marshes, names over a dozen species and indicates that there were more species present than he could name (Thesiger 2007, 114). Anderson (2007, 3) indicates that 'over a hundred species of birds, many of them migratory waterfowl [...] depended on the Marshes for refuge in the winter.' Wetlands are habitat for migratory waterbirds, including waders (as we have seen and will see on many occasions in the present volume). In June 1958 Thesiger (2007, 220) left the Marshes of southern Iraq never to return as three weeks later nationalist army officers led a coup in which the king, along with members of his family and the Iraqi prime minister were murdered in the so-called 'July 14 Revolution' (Anderson 2007, 7).

Gavin Young (2009, 9) sought to tell the story of 'what happened next' to the Marshes and the Marsh Arabs following the departure of Thesiger and Maxwell. He followed in Thesiger's footsteps in evoking the sights and sounds of the Marshes in similar terms (Young 2009, 21). Unlike Thesiger, he also evoked the smells of the Marshes (Young 2009, 17). Returning to the Marshes in 1973 Young (2009, 149) felt 'the old comfort in the deep grey-green womb of the reeds.' It gave birth to, and nurtured, a region where 'the Marshes are alive' and 'not a mere beautiful, stagnant backwater of the world' (Young 2009, 168, 123). He later wrote an 'Epilogue' in 1989 about 'the horror' and devastating 'meat-grinder of the war between Iran and Iraq' from 1980 to 1988 that inflicted the firepower of helicopter gunships, missiles and rockets on the Marshes and the Marsh Arabs (Young 2009, 171–178).

3.3 Marshland Cradle of Western Civilization

In 1968 Edward Ochsenschlager began conducting 'ethno-archaeological' fieldwork in the Marshes that he carried out until 1990. Ochsenschlager (2004, 1, 8–11) takes up the theme of the Garden of Eden as 'the name given to the "earthly paradise",' as 'an actual locality of earth' and links both together in the Marshes of southern Iraq which were 'a thing of beauty.' The account of the creation of the Garden of Eden in *Genesis* 2: 9 relates that 'out of the ground made the Lord God to grow every tree that is pleasant to the sight, and good for food.' Hence Woolley's conclusion of scarcely needing any labour to produce food from the Iraqi Marshes. According to *Genesis* 2:10, 'a river went out of Eden to water the garden.' Perhaps *two* rivers went *into* Eden to water the garden.

Rather than the biblical Garden of Eden, mud plays a greater role in Ochsenschlager's study of Iraq's Marsh Arabs. He relates how he was:

> brought up on a farm in Illinois where mud was considered a serious impediment to plowing, planting, harvesting, and any kind of transportation. I was totally unprepared to discover the many important uses of mud in the villages around al-Hiba [the site of his research in the Iraqi Marshes], where it was seemingly incorporated into every aspect of village life and architecture. (Ochsenschlager 2004, 45)

Ochsenschlager grew up with the typical mainstream western cultural and agricultural antipathy to mud. A traditional wetland culture, such as the Marsh Arabs, with its paludiculture had an entirely different view of mud and variety of uses for it. Sun-dried mud was made into drums, whistles, toys, jewellery and all kinds of cooking and storage vessels. Mud walls were also used in building construction.

Ochsenschlager goes on to give a chapter-by-chapter account of 'mud household utensils and storage containers' (Chap. 4), 'mud musical instruments, toys, jewelry [*sic*], and ammunition' (Chap. 5), 'mud architecture and ancillary structures' (Chap. 6) and 'baked pottery' (Chap. 7). He then turns to that other resource of the Marshes, reeds, and devotes a chapter to 'mats, baskets, and other objects made of reeds and rushes' (8) and to 'reed architecture' (9). More than mud and reeds were provided by the Marshes and Ochsenschlager goes on to discuss 'wood, boats and bitumen' (Chap. 10), water buffalo husbandry (Chap. 11), sheep and wool (Chap. 12) and 'village weaving' (Chap. 13). Ochsenschlager's first-hand, observational and detailed study of the lives of the Arabs in the Marshes interspersed with his own photographs in every chapter develop many of the aspects of Thesiger's account and add new ones to it.

In 1979 Saddam Hussein seized power in Iraq, started a war with Iran that ran from 1980 to 1988 and remained in power until 2003 with devastating consequences for the Marsh Arabs and the Marshes themselves. There were devastating consequences for researchers too, such as Jennifer Pournelle who was unable to visit Iraq to conduct fieldwork for her PhD Dissertation. She had to rely on and use aerial and satellite photography, archives and published sources, all of which she analyses thoroughly in several hundred pages of her dissertation (Pournelle 2003, xix, 22–259). She also contextualizes the western perception and study of Mesopotamian Marshes in the first, introductory chapter that 'examines a Western Enlightenment tradition hostile to uncultivated wetlands' (Pournelle 2003, xxxii), though the wetlands of southern Iraq were cultivated, just not in agriculture, but in paludiculture. She does not mention, nor discuss, any enlightenment philosophers, such as Immanuel Kant whose thinking on the sublime and the beautiful as aesthetic and cultural categories informs modern western antipathy towards wetlands as lacking these aesthetic qualities (see Giblett 1996, 30–31).

Pournelle (2003, 2–4) traces a brief and selective history of what she calls 'uncultivated wetlands' in England and the United States with various responses to coastal marshes and inland swamps depending on whether they were either in the city or the country. Wetlands in the country of both nations were cultivated by local, native peoples. They were later either usually drained to create dry farmland for agriculture, or they were used and conserved on farms for their resources of feed and water for grazing animals in paludiculture. By contrast, wetlands in and near cities in both nations were either drained to create dry urban land, or landscaped into lakes in parks and gardens (not mentioned by Pournelle). A much bigger and longer history of wetlands in England, the United States and other countries, and responses to them, has been told previously and will be retold in part in the present volume with references for further reading.

Pournelle traces the beginnings of 'the Western Enlightenment' hostility to 'uncultivated wetlands' back to what she sees as its first explicit expression in print in 1662 when the:

> English antiquarian and early modern historian Sir William Dugdale felt it necessary to defend the legitimacy of the reclamation practices [of English wetlands], in a *History of imbanking and draining of divers Fens and Marshes, both in foreign Parts and in this Kingdom, and of the Improvements thereby, extracted from Records, Manuscripts, and other authentic Testimonies.* The work was revised and corrected a century later by Charles Nalson Cole, and a second edition published in 1772. In it, Dugdale/Cole explicitly appealed to a series of proofs that equated embanking and drainage with great rulership. They argued that all the great biblical civilizations—Egypt, Babylon, Greece, and Rome—owed their wealth and prosperity to drainage works, and that the great English families had emulated this good work in the riparians of Britain. The Dugdale/Cole apology demarcates an Enlightenment turn that would dominate Western thinking regarding wetlands for the coming two centuries. Wedding the concepts of Biblical morality, urban civilization, technological innovation, and landscape transformation to productive increase (as defined by the urban landholder), coastal wetlands, perceived little more than a century earlier as a source of productive ease, became redefined as waste land awaiting the firm hand of good governance. (4–5)

Dugdale (1662, 2) not only invoked 'Biblical morality' in which wetlands were ostensibly bad and dry lands good, but also reached for his Bible for his opening gambit 'that works of Drayning are most antient, and of divine institution, we have the testimony of holy Scripture. In the beginning God said, let the waters be gathered together, and let the dry land appear.' Dugdale wilfully misquotes *Genesis* 1: 6 which states that

'God said, Let there be a firmament in the midst of the waters, and let it divide the waters from the waters.' Dugdale's 'Western Enlightenment' hostility to undrained wetlands appeals to a much older, middle-eastern scriptural tradition of Judeo-Christian theology to give it legitimacy. Dugdale syncretized Judeo-Christian theology with 'Western Enlightenment' views of development and drainage in which the engineer and the landscaper hostile to wetlands and with their wetland-modifying skills were the heroes of the age for their landowning employers. I discuss Dugdale's scriptural apology for the drainage of wetlands after the fact in greater detail in Chap. 4 of the present volume. I also discuss the drainage of the English Fens in Chaps. 4 and 5 of the present volume in which the drainage engineer Cornelius Vermuyden features prominently.

'The Western Enlightenment' hostility to 'uncultivated wetlands,' according to Pournelle, developed into 'those nineteenth-century beliefs that marshes are inherently diseased, sodden wastelands, and that the appropriate effort of good government was to transform them into cultivated agricultural land' (Pournelle 2003, 9). Yet these beliefs did not begin with Dugdale in 1662 in 'the Western Enlightenment,' or end in the nineteenth century (as we will see on many occasions in the present volume), but have their roots going much further back in time than that to ancient Greece which propounded the malarial theory of disease that diseases such as malaria were caught by breathing the bad air (literally 'mal-aria') arising from marshes. These beliefs also have their roots in ancient Rome that lauded cultivated agricultural land over uncultivated marshes (as we will see in the following two chapters of the present volume). These Greek and Roman views were syncretized with Judeo-Christian theology by Dante in the middle ages and John Milton in the English Renaissance and Reformation (as we will also see). Milton's *Paradise Lost* was published in the late seventeenth century after Dugdale's *History*. Both were on the cusp of the end of the English Renaissance and Reformation and at the beginning of 'the Western Enlightenment,' and situated in the tension and transition between them in the 1660s and 1670s.

'The Western Enlightenment' not only inherited the hostility to unreclaimed and undrained wetlands and other views about them via the Renaissance and its revival of ancient Greek and Roman learning, but also heightened them with its own racist, placist and sexist twist. 'The Western Enlightenment' arguably could not have got going without a corresponding and concomitant process of 'endarkenment,' whether it be of the black waters of wetlands that were denigrated ('blackened') literally and figuratively, or 'the dark continent' of Africa with its 'heart of darkness' for straight white male colonists in Joseph Conrad's novella of this title, or of female sexuality for straight white male Europeans according to Sigmund Freud. 'Western Enlightenment' and 'Endarkenment' go hand-in-hand; the former is not possible without the latter.

All of these views culminated in the first half of the twentieth century, beginning with the proto-Fascists (*Freikorps*) in Germany who rose to prominence in the aftermath of the first World War in the 1920s and who were the breeding ground for the Nazis in the 1930s. Pournelle (2003, 5 n6) relegates the gender politics of wetlands to a footnote that states 'in his now-classic exploration of early twentieth-century German masculinist ideologies, Klaus Theweleit dedicates an entire chapter to German biases against swamps.' Yet Theweleit*'s Male Fantasies* is about much more than mere masculinist biases against swamps that they feminised (which Pournelle neglects to mention). It is about out and out fascist hostility towards them. The fascist Benito Mussolini draining the Pontine Marshes in the mid-1930s is a prime example of masculinist hostility towards marshes (as we will see in the following chapter).

Fascism demonstrates hostility towards both mothering marshes and female bodies as a means of constituting the masculinist fascist body. Jessica Benjamin and Anson Rabinbach in their foreword to the English translation of volume two of Theweleit's *Male Fantasies* (1989) argue that:

> fascism is an extreme example of the political polarization of gender with two basic types of bodies [which] exemplify the corporeal metaphysics at the heart of the fascist perception: the soft, fluid, and ultimately liquid female body which is a quintessentially negative 'Other' lurking inside the male body [. . . and] the hard, organised, phallic body devoid of all internal viscera which finds its apotheosis in the machine. (cited by Giblett 1996, 45–46)

The swamp and the body of the mother are the soft, fluid, and ultimately liquid body *par excellence* for the fascist soldier machinic male. Barbara Ehrenreich in her foreword to the English translation of volume one of Theweleit*'s Male Fantasies* (1987) concludes that 'for the fascist soldier male, "women's bodies are the holes, swamps, pits of muck that can engulf"' (cited by Giblett 1996, 45). Conversely, many men have feminized and figured swamps as the holes and pits of muck of women's bodies that can engulf the unsuspecting and unwary straight white male.[1]

These misogynist views about women's bodies culminated later in the first half of the twentieth century in Jean-Paul Sartre's *magnum opus, Being and Nothingness,* published in French in 1943. It is purportedly one of the greatest philosophical works of the twentieth century. It is certainly one of the most misogynist works of the twentieth century. In it he devotes ten markedly misogynist pages to a philosophical meditation on what he calls 'the horror of the slimy' (cited by Giblett 1996, 40). For Sartre, the slimy is feminine and an object of horror for what he calls 'the European adult' (cited by Giblett 1996, 40). Swamps are the slimy place *par excellence* and a feminized object of horror for the straight white male adult who is constituted as such by *not* being the feminized slimy swamp and by being horrified by the (s)mothering body of the swampy mother and mothering swamp.[2] The Existentialist Marxist Sartre and the fascist soldier male seem to share a common gender politics. 'Western Enlightenment/ Endarkenment' bias towards undrained and unreclaimed wetlands is more precisely masculinist and misogynist hostility towards them.

Beginning with Aldo Leopold's watershed *Sand County Almanac* (1949), in the second half of the twentieth century this hostility has not only been critiqued, deconstructed and decolonised (such as by Franz Fanon in 1961 as we saw in the introduction to the present volume), but has also been countered with feminist, post-colonial and conservationist views and appreciation for wetlands. All of these aspects, approaches and politics were brought together at the end of the twentieth century in *Postmodern Wetlands* that is now regarded as a pioneering work in wetland cultural studies and a classic in the transdisciplinary environmental humanities, along with *Western Australian Wetlands* published in the same year (Giblett 1996; Giblett and Webb 1996). These aspects, approaches and politics in relation to wetlands continue to be developed in the twenty first century in subsequent studies (Giblett 2021).

[1] Several instances spring readily to mind, including H. Rider Haggard in his adventure romance, *She* (see Giblett 1996, 131), and Joseph Conrad in his first published short story, 'The Lagoon' (see Giblett 1996, 96–97).

[2] For further discussion of Sartre on slime, see Giblett (1996, 39–47), and of the mothering marsh, see Giblett (1996, 145–150)

3.4 Muck, Mire and Reeds

Muck and mire crop up in Pournelle's dissertation when she discusses *Zenaide* Ragozin's *Chaldea: From the Earliest Times to the Rise of Assyria* first published in 1886 and went through several editions. I cite the fourth edition of 1893 readily available online. Her book was later republished as a volume in Hutchinson's school textbook series *Story of the Nations* under the title of *The Story of Chaldea*. It is still in print. Pournelle (2003, 10) states that it was 'widely read for decades in multiple editions as a general introduction to the study of ancient history.' She goes on to argue that Ragozin's:

> colorful illustration of pre-literate, pre-civilized Sumerians anchoring, behind reed-and-earth dams, baskets of muck dredged from a waste-deep mire remained imbedded in research paradigms through the mid-twentieth century that viewed the birth of Mesopotamian civilization as a process of "reclaiming" primordial lands for irrigated plow agriculture. (Pournelle 2003, 10)

These research paradigms are also mired in the mainstream western cultural paradigm that denigrated wetlands and advocated 'reclamation' of them either for irrigated agriculture, or for dry land agriculture. 'Reclaiming' implies that wetlands stole the land for water and that the land needed to be reclaimed (or wrested back) from water by draining wetlands. Through the mid-twentieth century these paradigms were being countered by progressive, alternative and counter-cultural paradigms that appreciated wetlands and advocated their conservation and/or rehabilitation championed by Leopold and many others.[3]

Although Ragozin does not use the words 'muck' or 'mire,' she does subscribe to the mainstream miasmatic theory of disease in the west. She claims that 'the lowlands towards the Persian Gulf' present an 'inconvenience' where 'the waters accumulate and stagnate forming vast pestilential swamps where pastures and wheat-fields should be' (Ragozin 1893, 7). Later she goes on to describe 'the unhealthy exhalations of a vast tract of swamps, and […] the intolerable nuisance of swarms of aggressive and venomous insects, which infest the entire river region' (Ragozin 1893,

[3] See for instance, John and Mildred Teal's *Life and Death of the Salt Marsh* (1969), published twenty years after Leopold's *A Sand County Almanac*. For the mainstream and alternative cultural paradigms presented in tabular form, see Giblett (2011, Chap. 1, Fig. 2).

46). Ragozin is singing straight from the song-sheet of the miasmatic theory of disease first propounded by the ancient Greeks (as we will see in the following chapter of the present volume). The miasmatic theory of malaria was superseded by the contagion theory when Ronald Ross established the vector for malaria in 1892. The news does not seemed to have reached Ragozin in San Antonio in New Mexico where she was living at the time. She is also repeating the biblical interdiction against swarming creatures as an abomination (*Leviticus* 11: 41).

Against the mainstream 'Western Enlightenment' hostility towards 'uncultivated wetlands,' Pournelle (2003, 18) 'hypothesize[s] the essential nature, not merely of *water*, but of *wetlands*, in supporting and shaping the complex social institutions that underlay urbanization in southern Mesopotamia' (her emphases). Pournelle takes up this hypothesis in Chap. 4 of her thesis where she argues that:

> a significant component of the resource basis for precocious, large deltaic towns was probably that derived from surrounding marshland; and conclude[s] that only following specialization and integration of, not two, but three, productive economies: agricultural, pastoral, and wetland, could and did Mesopotamian urban civilization flourish. (Pournelle 2003, xxxii, 20)

The wetland productive economy is what I call 'paludicultural' in which wetlands were cultivated. Pournelle amasses and analyses a huge amount of topographical data derived largely from aerial and satellite photography of southern Mesopotamia in support of her hypothesis and conclusion that wetlands were cultivated.

In particular, Pournelle counters 'the two great rivers' (213) theory of Mesopotamia perpetuated by Maxwell in which mainstream western culture and conventional anthropology and archaeology privileges water flows between clearly defined banks (rivers, such as the Tigris and Euphrates, and embanked wetlands in forming rivers and irrigation channels) and relegates wetlands to the secondary status of being between the rivers and a product of them (literally 'Mesopotamia,' between rivers). Pournelle foregrounds the importance of wetlands when she contends that what she calls 'the lower alluvium' (212) of southern Mesopotamia was:

> inundated by multiple distributary fans that formed wetlands such as those of the modern inner (freshwater) and outer (brackish–saltwater) deltas. (Pournelle 2003, 213)

What have been called rivers are the outlying eastern and western extent of these deltaic wetlands. The two great rivers are outgrowths of, and adjuncts to, the great wetland.

Pournelle's counter-theory is what could be called the contemporary 'one great wetland' theory of southern Iraq—not Mesopotamia (between rivers), but Megapalus, the great marsh. It is akin to what I have called 'Middlemarsh,' or Mesopalus, with its wetlands and fertile plains between 'the Grampians' (Gariwerd) mountain ranges and the Hopkins River in south-eastern Australia. With its wetlands, fertile plains and 'the Rainbow Serpent' of the Hopkins River in Aboriginal story, Mesopalus is a fertile serpent akin to the fertile crescent of Megapalus. The first European explorer to see Mesopalus called it the Garden of Eden, though he may have had John Milton's Eden in *Paradise Lost* more in mind than the Biblical one, though Milton does not include marshes in his description (Giblett 2023, 83, 205–207n37). Nevertheless, Mesopalus is akin to Megapalus as the Garden of Eden too. Like Megapalus, Mesopalus is a cradle of civilizations too (Giblett 2023, 23–27).

The final chapter of Pournelle's dissertation concludes by contrasting the modern, 'Western Enlightenment' and colonial view of wetlands with the way:

> wetlands may have been seen by proto-urban elites five millennia earlier, as they undertook their urbanizing projects that eventually converted socially unranked, undifferentiated wetlands, into alienated, ranked, extra-urban hinterlands.

In support of this view, Pournelle (2003, 219) chooses to focus on:

> one feature of the Iraqi wetland terrain—the ubiquitous reed—that, even in the face of overwhelming evidence of its cultural and social importance through time has remained largely invisible to those who could not or did not recognize the real significance of wetlands or the (economic, social, or cultural) value of "non-agricultural" products or communities.

Young, Maxwell, Thesiger and Ochsenschlager recognize and document the cultural and social importance of reeds as a resource, but do not link them to their cultivation in the Marshes and to the Marsh Arabs' productive economy of paludiculture. As Pournelle (2003, 219) maintains:

> reeds were themselves at all times a valuable resource; an ecological component of the agro-pastoral[-paludal] system that have contributed significantly to rural surplus generation in the service of urban populations [...] In all of [the distinct zones of the wetlands of the southern delta] the predominant activity is reed-cutting for construction, fodder, fuel, basketry and reed matting produced for barter or sale. (240)

Ochsenschlager considers the use of reeds for all of these activities, but discusses them in isolation in an inventory of cultural practices and artefacts, not for their central and vital role in the wetland productive economy and as the foundation for cities discussed by Pournelle (who he cites, but not on these points). The bigger picture of what is at stake here historically, environmentally and culturally gets lost in the attention to, and study of, details—a plain enough case of not being able to see the wood for the trees.

Reeds are a valuable resource for sustaining the Marshes and the Marsh Arabs.

Lonergan and Al-Asadi (2024, 14) relate that:

> The reeds in the Marshes were almost as important as the water to residents. It wasn't only houses that were built from reeds; so, too, were the shelters and fences used to house and constrain animals, and the mats used for sitting and sleeping. Even the school was built from reeds.

Reeds need water, but water alone cannot sustain humans and water buffalo, who sustain human life (Lonergan and Al-Asadi 2024, 114–115). Reeds, water and land make up the Marshes. This was where Jassim Al-Asadi was born in 1957. He and Steve Lonergan tell his story of how he 'grew up in the Marshes, and they became part of his heart and soul [...] The boy was born in nature and would forever be part of nature' (Lonergan and Al-Asadi 2024, 13, 15). They interweave his story with the history and politics of the Marshes and the Marsh Arabs from ancient times and Mesopotamian myth to the early 2020s and recent conservation efforts.

3.5 Conservation of Iraq's Marshes

The paludicultural productive economy was largely destroyed under Saddam Hussein's regime. In his introduction to Thesiger's *Marsh Arabs,* Anderson relates the sad and sorry story of the Marshes after Thesiger left in 1958, culminating in Saddam Hussein's reign of terror in the 1980s and 90s largely destroying the Marshes, but ending on a brighter, recent note of wetland rehabilitation. During the Iran-Iraq War of 1980–1988 the Marshes were a sanctuary for 'a large number of deserters from the Iraqi army.' After the Gulf War of 1991, Shiite rebels began a revolt in Basra and 'kept up a low level insurgency, using what was left of the Marshes as a sanctuary' (Anderson 2007, 7–8). Marshes and other wetlands have been a sanctuary for a long time before then for deserters and draft dodgers, such as John Muir during the American Civil War (1861–1865), a site for resistance for rebels against empire, such as the Fenlanders (as we will see in later chapters of the present volume) and Francis Marion, 'the Swamp Fox,' in the American Revolutionary War (1775–1783), and a refuge for runaways, such as slaves in the American south in the nineteenth century.[4] Lonergan and Al-Asadi (2024, 76) relate that:

> There are many reasons why armies avoided fighting in the Marshes. While mythical demons ('Afarit) and supernatural creatures (jinn) may be among them, the difficulty in traversing the mud, reeds, and water, the impossibility of bringing horses into the wetlands, and the ability of the residents of the Marshes to foment guerrilla warfare and then disappear, all made conquering the region virtually impossible. The tribes of the south were formidable adversaries—better to stay on the edge of the Marsh and let well enough alone. (Lonergan and Al-Asadi 2024, 76)

Yet Iran and Iraq could not leave the Marsh alone during their war of the 1980s when the 'marshes and wetlands were a major impediment to an advancing Iranian army, but they could pose transportation problems for Iraqi forces as well' (Lonergan and Al-Asadi 2024, 77). The Marshes

[4] For further discussion of wetlands as a site of resistance for rebels and patriots (including Francis Marion, 'the Swamp Fox,' during the American Revolutionary War) and a refuge for runaways (such as slaves in nineteenth-century America), see Giblett (1996, Chap. 9). For further discussion of John Muir, the founder of the Sierra Club and its first President, and his sojourn in a Canadian marsh with deserters and other draft dodgers from the American Civil War, see Giblett (2014, Chap. 7).

exemplified the unique role of wetlands in warfare of being easy to defend, but hard to attack as theorized by von Clausewitz in the eighteenth century (Giblett 1996, xx–205).

After the Iraqi victory of 1988, 'what had been a war *in* the Marshes soon developed into a war *on* the Marshes' (Lonergan and Al-Asadi 2024, 86–87). Destroying Iraq's Marshes as a sanctuary and its marsh dwellers was the impetus behind:

> the 1992 decision to drain the Marshes [...] based on the Iraqi regime's desire to repress the local population, transform an area that had served to shelter those fleeing political and religious persecution, and facilitate oil and agricultural development. To these ends, the government bombed and razed villages, resettled the local population to towns and cities, and purposely dried up over 90 percent of the wetlands. (Lonergan and Al-Asadi 2024, 217)

Lonergan and Al-Asadi (2024, 3) relate that drainage and dispossession were undertaken:

> ostensibly to reclaim land for agricultural development and promote economic modernization, [but they] almost destroyed the Marshes [in] one of the greatest environmental and humanitarian disasters of the twentieth century.

Anderson relates that draining the Marshes involved:

> the rapid construction of an immense canal to drain the area. When the canal was completed six months later, it was named Saddam River. By 1994, as much as 90% of the Marshes had been destroyed, and most of the Marsh dwellers had been killed, or had fled into exile in Iran, or moved to the slums of Basra and other Iraqi cities. What had been pristine Marshlands teeming with birdlife and water buffaloes had become the vast baked pan of an empty desert. (Anderson 2007, 8; see also Lonergan and Al-Asadi 2024, 109–112, 117–119)

In 2003 after Saddam Hussein was toppled (including his statue) during the invasion of Iraq by 'Coalition' forces Iraq's Marshes:

> began a revival. As soon as Saddam Hussein had fled Baghdad in April 2003, tribesmen took shovels and diggers, and tore holes out of the sides of the

> earthen dikes that Saddam's engineers had erected. Water poured through, and began to flood over the barren land. Within a year, an estimated 20% of the former Marshlands had been re-flooded. Iraq's government began working with a consortium of international experts to restore the Marshes, but scientists were divided over how much of the original Marshlands could be revived. Some said as much as 80%, other said only a third. In some places vegetation began to reappear, while other areas remained barren. Some of the displaced Marsh Arabs returned to rebuild their villages, but many more did not. (Anderson 2007, 10–11; see also Lonergan and Al-Asadi 2024, 138–143)

The plight of the Marshes became international news. Jassim Al-Asadi became the senior technical engineer of the Center for the Restoration of the Iraqi Marshes (CRIM) (Lonergan and Al-Asadi 2024, 136) and more besides:

> Jassim was slowly becoming a leading advocate for the preservation of the Marshes, both nationally and internationally. As part of his work for the CRIM, he carried out engineering and environmental assessment projects and worked closely with local communities in and around the Marshes. He gave lectures on the Marshes at national and international conferences, conducted archeological surveys of the area, organized workshops, and promoted environment and development projects. Jassim also increased his cooperation with Nature Iraq, edited a column titled "A Window on the Environment" in a Baghdad newspaper, and supported the development of veterinary centers for Marsh dwellers. Restoring the Marshes, however, required more than additional water and engineering structures; it needed marketing. This involved convincing both the Iraqi people and government agencies that the Marshes were a unique resource worth protecting. Engaging the international community was part of this, both in terms of receiving necessary financial assistance and recognizing the importance of the Marshes and the people living there. Most of all, an effort was needed to help people living in the Marshes understand the changes taking place, and to support them by advocating for improved health and education services; Nature Iraq was crucial for both. (Lonergan and Al-Asadi 2024, 154)

Nature Iraq was also crucial for:

> increasing awareness of the Marshes through the UN sustainable development conferences in 2004 and 2005. The Ministry of Environment, CRIM, and Nature Iraq soon set their sights on having Hawizeh Marsh designated

> as a wetland of international importance under the Ramsar Convention. There are over 2,300 Ramsar sites around the world, and by 2007 Iran had twenty-one designated sites, while Iraq had none. To become a member of the Ramsar Convention, a country must designate at least one wetland as a potential site. This site must meet one or more of nine criteria set forth in the Convention. Once it becomes a member of Ramsar, the country is then obliged to manage the wetland to ensure that ecological communities are maintained. The issue of gaining Ramsar designation was important enough that CRIM changed their name to CRIMW, the Center for Restoration of Iraqi Marshes and Wetlands, in 2007. The northern section of Hawizeh Marsh, one of the few wetlands that withstood the devastation of the 1990s unscathed, met five of the nine Ramsar criteria and was placed on the List of Wetlands of International Importance in 2007. (Lonergan and Al-Asadi 2024, 155)

As well as a section of the Iraq Marshes becoming a World Heritage Site in 2016:

> in November 2022, the city of Chibayish was named a Global Wetlands City at the UN's Convention on Wetlands—one of only twenty-five cities that have received this designation. The reason is partly that Chibayish sits on a 300-meter strip of land between the Central Marsh to the north and West Hammar Marsh to the south. Other criteria for being awarded the designation include: ensuring development does not degrade the wetlands, that the wetlands play a role in enhancing human well-being, the inclusion of indigenous and local communities in decision-making, and promoting events around World Wetlands Day. The UN designation is intended to raise public awareness of the benefits of wetlands and encourage wetlands protection in municipal planning. (Lonergan and Al-Asadi 2024, 168)

The only North American city to achieve this designation is Sackville in the Canadian province of New Brunswick set beside the Tantramar Marshes. None of the other North American cities set in or beside wetlands, such as Boston, New York, Toronto, Chicago, Vancouver, New Orleans and Washington have achieved it. No Australian city, such as Perth Melbourne, Cairns and Mackay, has achieved it either, despite being set in or beside wetlands.

Lonergan and Al-Asadi outline the:

> reasons why the Marshes have been internationally recognized as worth preserving. Wetland ecosystems are essential in storing and treating water,

fostering biodiversity, and moderating climate change, while water reflects solar radiation and plants absorb carbon dioxide. The result is a regional cooling effect. Dry land absorbs more solar radiation and makes a region hotter, and a lack of water can also cause dust storms—as happened in the mid-1990s—with impacts on the health of both humans and animals. Wetlands are natural filtering systems for agricultural, industrial, and human waste. The Marshes have been effectively treating waste for thousands of years. It is only when waste loads became too large that problems arose [...] In addition, wetlands are contributors to socioeconomic development. The Marshes continue to support extensive numbers of fish, migratory and resident birds, amphibians, and plants. [...] Sustained national development is increasingly being recognized by the international community as having three components: economic, social/cultural, and environmental. The Marshes embody all three of these components, countering the perception that the wetlands are an undeserving recipient of valuable fresh water from the north. In addition, the Marshes play a crucial role in engaging Iraq with the international environmental community. This is no small achievement, as many countries continue to view Iraq with suspicion due to the high level of governmental/private sector corruption and the uncertain security situation. In a water-scarce region, every drop is important—for drinking and cooking, agriculture and industry, energy production, and transportation and recreation. Water is also essential to support ecosystem functions which, in turn, support economic activities such as fishing. The problem is that when fresh water is scarce, it increases the likelihood of conflict among these different uses. Much can be accomplished to prevent this. The agricultural sector is a place to start: improving the irrigation infrastructure and phasing out water-intensive crops, such as rice, would provide both financial benefits and water savings. Enhancing the efficiency of water use by industry is another option; for instance, fresh water should not be injected into wells to support oil reservoirs. Moreover, treatment systems need to be improved. The Marshes are not the culprit in the battle for water, but rather should be viewed as part of the solution. Admittedly, it is no easy task to convince everyone that the Marshes are important to preserve for reasons of culture, biodiversity, or even ecosystem function—especially in the face of poverty, unemployment, and internal conflict and strife. This is why it is important to have more individuals and groups who are committed to preserving the Marshes and whose voices will be heard at the highest levels of government and in international forums. There is a dire need for advocates for a change in the way Iraq, and, indeed, all countries, balance the importance of nature with powerful economic interests. (Lonergan and Al-Asadi 2024, 212–215)

The Marshes also have great longevity and an important place in human history, as Lonergan and Al-Asadi go on to relate:

> The Iraqi Marshes formed over 10,000 years ago and have demonstrated remarkable resilience over the millennia. [...] Human presence in the Marshes dates back at least 7,000 years. The first city-state, Uruk, was settled in 4500 BCE. From that time onward, the Marshes became more than a unique ecological area; they developed into an exceptional social, economic, and cultural region as well. The existence of fresh water in southern Mesopotamia allowed human civilization to flourish. The wetlands stored and purified the water and acted as an effective wastewater treatment system, filtering organic matter and nutrients. They were home to a wide variety of plants and animals, fostering biodiversity and helping sustain human life. (Lonergan and Al-Asadi 2024, 216–217)

Over the past 20 years:

> local initiatives helped to partially reflood the Marshes. Dense reed patches returned, although they were less widespread than in the 1950s. Migratory and resident bird populations rose to 1970s levels. An active Marsh culture remained, although it was much reduced in size. Fishing again became viable, even though the dominant species changed, with an exotic species of carp now accounting for most of the catch. Buffalo numbers are much lower than before the draining, although the remaining animals still wander deliberately into the marshes in the morning and return to their islands late in the day. Their languorous movement may not be in harmony with motorboats, but modernity and the past seem to have reached an acceptable coexistence—except for in years of extreme drought. Depending on the year and the season, the wetlands now vary between 25 and 50 percent of their size in the early 1970s, with half of this being temporary marsh. Most notably, there have been measurable reductions in water quality, biodiversity, and ecological productivity. And while the Marshes have shown remarkable resilience, they continue to be vulnerable to upstream development and climatic change. Severe drought in 2009, 2015, 2018, and 2021–2022 dried out many shallow wetland areas and increased the concentration of pollutants in the remaining water. (Lonergan and Al-Asadi 2024, 218)

Al-Asadi has lived through these changes to the Marshes in the past from the 1950s when he was born to the present of the 2020s when he still works for the Marshes.

The Marshes face an uncertain future for:

> Despite their impressive resilience, the Marshes are under threat—of this, there is little doubt. The environment Jassim knew as a boy has been ravaged by social, political, and economic upheaval. Reduced volumes of water in the rivers from upstream dams, agricultural runoff, urban wastewater, large-scale oil development, and a political culture that undervalues preservation of the Marshes combine to offer a bleak picture of their future. The reality of climate change, of drought that can cripple both the wetlands and the people living nearby, casts a pall over not just the region, but the entire globe. The wetlands are now fragmented, and the annual pulse of floodwater that once provided needed fresh water and cleansed the Marshes of pollutants is gone. Today, few people live within the Marshes. Technological change has impacted the region, as it has globally. Shakhtura have all but replaced the traditional *mash-huf* [a bitumen-covered wooden canoe for two persons that is used for general transportation, fishing, and collecting reeds]; aquaculture in the rivers and electric shock fishing are more common than net fishing; and sheikhs no longer use the graceful tarada for transportation, but instead use modern trucks and automobiles. (Lonergan and Al-Asadi 2024, 219, 239)

Al-Asadi took early retirement from his government position as an irrigation engineer and he is now the head of the non-government organization, 'Nature Iraq,' which means that he is a full-time environmental activist. It is a dangerous cause and occupation in Iraq. When he was a university student in 1978, he was kidnapped and tortured. Forty-five years later history repeated itself. He was kidnapped on 1 February 2023 as he drove to Baghdad to meet the President of Iraq. Steve Lonergan relates in an 'Afterword' to *The Ghosts of Iraq's Marshes* that:

> It should be no surprise that being an environmental activist in Iraq, a public advocate for water use and water rights, and a champion for ensuring the Marshes receive enough water to survive, carries a high degree of personal risk. On February 1, 2023, Jassim was kidnapped while driving with his cousin Raad from Hilla to Baghdad for a meeting with the president of Iraq. Two cars blocked the road, and five men approached Jassim's car with their guns drawn. They pulled Jassim out of the car, covered his face, gagged him, tied his hands, and dragged him into another vehicle. He was shoved roughly between two seats, and the two cars quickly departed. (Lonergan and Al-Asadi 2024, 235)

Following intense local and international pressure, he was released on February 16. On February 17 he:

> revealed the details of the crime of his kidnapping, torture and abuse for 15 days, in the activist's first comment after his release. During his detention, he was subjected to torture using electricity and sticks. He suggested that the purpose of his kidnapping was to dissuade him from continuing to work for the marshes to perpetuate this culturally and economically important environment for Iraq. (Anon 2023)

Lonergan concludes that:

> Despite his ordeal, Jassim remained resolute. "The support I have received . . . makes me stick to the cause for which I was kidnapped, and I will continue my same path of defending the environment [meaning the Marshes] and its inhabitants," [...] Jassim is adamant that he will remain in Iraq and work for the preservation of the Marshes and the welfare of its people. As he told me, "Jassim without the Marshes is not Jassim" (Lonergan and Al-Asadi 2024, 238).

Like the wetland Aboriginals of Queensland, for the Marsh Arabs of Iraq wetlands mean life.

References

Anderson, Jon Lee. 2007. Introduction. In *The Marsh Arabs*, ed. Wilfred Thesiger, 1–11. London: Penguin.

Anon. 2023. The First Statement of Activist Jassim Al-Asadi After his Release. *ALShaqiya News*. https://www.alsharqiya.com/en/news/the-first-statement-of-activist-jassim-Al-Asadi-after-his-release. Accessed 8 June 2023.

Dugdale, William. 1662. *The History of Imbanking and Drayning of Divers Fenns and Marshes, both in Forein Parts and in this Kingdom, and of the Improvements thereby Extracted from Records, Manuscripts, and Other Authentick Testimonies*. Accessed September 16, 2022. https://quod.lib.umich.edu/e/eebo/A36795.0001.001?view=toc.

Giblett, Rod. 1996. *Postmodern Wetlands: Culture, History, Ecology*. Edinburgh: Edinburgh University Press.

———. 2011. *People and Places of Nature and Culture*. Bristol: Intellect Books.

———. 2014. *Canadian Wetlands: Places and People*. Bristol: Intellect Books.

———. 2016. *Cities and Wetlands: The Return of the Repressed in Nature and Culture*. London: Bloomsbury.

———. 2021. *Wetlands and Western Cultures: Denigration to Conservation*. Lanham, MD: Lexington Books.

———. 2023. *Middlemarsh: The Hopkins River, Kindred Wetlands and Remarkable People*. London: Transnational Press.

Giblett, Rod, and Hugh Webb, eds. 1996. *Western Australian Wetlands: The Kimberley and South-West.* Perth, Western Australia: Black Swan Press/ Wetlands Conservation Society.

Leopold, Aldo. 1949. *A Sand County Almanac and Sketches Here and There.* Oxford: Oxford University Press.

Lonergan, Steve, and Jassim Al-Asadi. 2024. *The Ghosts of Iraq's Marshes: A History of Conflict, Tragedy, and Restoration.* Cairo, Egypt and New York: The American University in Cairo Press.

Maxwell, Gavin. 1966. *People of the Reeds.* New York: Pyramid Books. First published 1957.

O'Gorman, Emily. 2021. *Wetlands in a Dry Land: More-Than-Human-Histories of the Murray-Darling Basin.* Seattle, WA: University of Washington Press.

Ochsenschlager, Edward. 2004. *Iraq's Marsh Arabs in the Garden of Eden.* Philadelphia, PA: University of Pennsylvania Museum of Archaeology and Anthropology.

Pournelle, Jennifer. 2003. Marshland of Cities: Deltaic Landscapes and the Evolution of Early Mesopotamian Civilization. PhD Diss., San Diego: University of California. https://www.academia.edu/843377/Marshland_of_Cities_Deltaic_Landscapes_and_the_Evolution_of_Early_Mesopotamian_Civilization. Accessed 27 Apr 2023.

Ragozin, Zénaïde A. 1893. *Chaldea: From the Earliest Times to the Rise of Assyria.* 4th ed. London and New York: T. Fisher Unwin and G. P. Putnam's Sons. https://www.gutenberg.org/files/24654/24654-h/24654-h.htm. Accessed 8 June 2023.

Scott, James C. 2017. *Against the Grain: A Deep History of the Earliest States.* New Haven, CT: Yale University Press.

Teal, John and Mildred. 1969. *Life and Death of the Salt Marsh.* Boston, MA: Little, Brown.

Thesiger, Wilfred. 1958. Marsh Dwellers of Southern Iraq. *National Geographic* 113 (2 (February)): 204–239.

———. 2007. *The Marsh Arabs.* London: Penguin. First published 1964.

Theweleit, Klaus. 1987. *Male Fantasies, Volume I: Women, Floods, Bodies, History.* Trans. S. Conway. Cambridge: Polity Press.

———. 1989. *Male Fantasies, Volume II, Male Bodies: Psychoanalyzing the White Terror.* Trans. C. Turner and E. Carter. Cambridge: Polity Press.

Woolley, C. Leonard. 1965. *The Sumerians.* New York: W. W. Norton. First published 1928.

Young, Gavin. 1976. Marsh Arabs, Water Dwellers in a Desert World. *National Geographic* 149 (4 (April)): 502–523.

———. 2009. *Return to the Marshes: Life with the Marsh Arabs of Iraq.* London: Faber. First published 1977.

CHAPTER 4

Malarial Greeks and Their Dispersion

The denigration of wetlands in mainstream western culture arguably began with ancient Greek philosophy and with the Hippocratic writings on medicine. Wetlands were regarded as the source of disease when miasmatic vapours rise from stagnant pools of water and cause malaria (literally 'bad air'). The miasmatic theory of disease prevailed up until the 1890s CE. It was based on the Greek philosophy of the four elements: earth, air, fire and water. Wetlands are problematic, indeed aberrant, from this point of view because they mix the elements of air and water in miasma and malaria (as well as air and 'fire' or heat in the humid tropics). These elements should remain separate in the Greek view. They should stay in their proper places in the clean, disease-free body of both the earth and human beings. They also cross the boundaries between land and water, and can even be in transition spatially and temporally between open water to dry land. They are a troubling and unsettling category between land and water mediating them.

As wetlands mix the elements, they produce an aberrant 'humour,' or psychosomatic state, strictly a kind of phlegmatic melancholy. In mainstream western culture, wetlands have been seen as a wilderness to be tamed, the sites of mixed elements and aberrant humours giving rise to melancholy and madness. They are also the home of monsters to be killed by heroes, such as Hercules killing the Stymphalian birds. Miasmatic

R. Giblett, *Wetland Cultures*, Literatures, Cultures, and the Environment, https://doi.org/10.1007/978-3-031-57365-1_4

marshes give rise not only to malaria and melancholia, but also to monsters and the monstrous. For Homer in *The Odyssey* wetlands are the underworld of Hades. Virgil, Dante and Milton followed suit with hell portrayed as a swamp (as we will see in the following chapter of the present volume).

Ancient Greek wetland culture of malaria, miasma, monsters and melancholia was dispersed to England and its wetlands, especially the Fens. Mainstream English wetland culture prescribed drainage of the Fens as the cure for all of these maladies. The same diagnosis underpinned the Sanitary Movement of the nineteenth century where the polluted, stagnant waters of cities were figured as miasmatic swamps (as we will see later in the present chapter).

4.1 The Elements

The theory of the elements was developed by the pre-Socratic philosophers and later enunciated by Aristotle in the treatise which came to be called 'De Generatione et Corruptione' ('On Coming-to-be and Passing-Away'). For Aristotle (cited by Ross 1995, 104) what he called the four so-called elements of earth, air, fire and water were produced by the selective 'coupling' of the four 'elementary qualities' and two 'contrarities' of hot and cold, dry and wet (or 'moist' as the translator prefers, or 'fluid' as one commentator prefers). As a result earth is cold and dry, air hot and moist, fire hot and dry and water cold and moist. Theoretically it was possible for the four 'elementary qualities' and two 'contrarities' of hot, cold, dry and wet to be combined into six couples, but as it is impossible for something to be both hot and cold, wet and dry (though not in practice for a solid to be damp or sodden as in a slimy wetland, nor for a solid to be sublim(at)ed into a gas by heat), there were in effect only four proper couples. It was thus acceptable for the elementary and contrary qualities to mix, but only in four prescribed or proper ways to produce the four elements. The mixing of the 'contrarieties' of hot and cold, wet and dry was 'impossible,' or in David Ross' (1995, 105) terms they 'refuse to be coupled,' though Aristotle (1930, §§329a–332b) later concedes that 'all the 'elements' will be able to be transformed out of one another' which would involve coupling of the contrarities.

Wetlands were effectively ruled out of court in this cosmogony on two counts depending on whether they were considered at the level of elements or qualities. They were ruled out at the level of elements because they combine earth and water, or because they mediate between earth and water spatially or temporally. They were also ruled out at the level of elementary qualities for as earth is cold and dry, and water cold and wet, a mixture of or a mediating category between the two sets of contrarieties would produce the quality of cold, dry and wet, an evident impossibility in this philosophical schema. The sublime is even more out of court as it involves a transition through the elements from 'earth' (or solid) to 'air' (or gas) by or via 'fire' (or heat). It would also entail the absurd combination of the qualities of hot, cold, dry and wet.

Aristotle distinguishes the elementary qualities of the hot and cold from the wet and dry in terms of the active/passive distinction: 'the first pair implies *power to act* and the second pair *susceptibility*' (§329b). Or as David Ross (1995, 105) put it, 'hot-cold play in general the part of agent and dry-fluid that of patient.' The distinction between agent and patient is gendered in patriarchy. The former is associated with virile masculine activity and the latter with second supine feminine passivity. Aristotelian philosophy is no exception as it, according to Carolyn Merchant (1980, 13), 'associated activity with maleness and passivity with femaleness.' On the Aristotelian view of (re)production, the masculine hot-cold inseminates the feminine wet-dry and brings forth the four elements. A gender distinction also applies to the elements. Of the four elements for Aristotle 'Fire and Earth [...] are extremes and purest' and 'Water and Air [...] are intermediaries and more like blends' (§§330b–331a). In other words, fire and earth are definite and discrete, and 'masculine' in patriarchal terms; water and air are mediatory and mixed, and 'feminine' in patriarchal terms. Fire and earth are privileged over air and water and hot and cold are over wet and dry; so the figure below is hierarchical.

Yet Aristotle accepted that 'Earth has no power of cohesion without the moist' (presumably just as long as the moist stayed between the bits of earth and did not mix with them thus dampening them) (§335a). 'Feminine' moist (the wetlands), holds 'masculine' earth (the dry lands) together; without the former the latter would fall apart. Wetlands are, as an article in *National Geographic* argues, 'the glue that holds the land together' (Mitchell 1992, 15). More precisely, wetlands are the soft glue that holds, in the words of Liam Davison's *Soundings*, 'the hard bones of the land' together (Davison 1993, 85). The moist, moreover for Aristotle, is 'that which being readily adaptable in shape is not determined by any limit of its own' (§329b). This 'feminine' adaptability and limitlessness of water is the condition of possibility for modern 'masculine' waterworks.

Yet while Aristotle was appreciative of the cohesive and other qualities of the moist, he certainly did not like the inundatory or penetrative qualities of the damp and the sodden which result from the solid and the liquid, earth and water, the dry and the wet, the masculine and the feminine 'impossibly' combining in aberrant ways: 'damp' is that which has foreign moisture on its surface ('sodden' being that which is penetrated to its core)' (§330b). In Aristotelian terms a wetland would be earth with 'foreign' moisture on it, as in a lagoon, or earth 'penetrated' with moisture to its core, as in a marsh. Either way, 'masculine' earth should not have 'feminine' water lying on its surface or penetrating its depths. The only proper function for 'feminine' water is to make the 'masculine' earth stick together. On this Aristotelian view and by implication, earth should not have the 'foreign' matter of water on it, nor should it be 'penetrated' by the invasive power of water. Earth should be dry for Aristotle and not wet.

4.2 Unhealthy Waters

Probably the earliest extant discussion of the association between wetlands and disease in western culture is in the Hippocratic writings. In the essay 'Airs, Waters, Places,' the Hippocratic writer is concerned to distinguish 'the effect of different kinds of water, to indicate which are healthy and which unhealthy, and what effects, both good and bad, they may be expected to produce' (Lloyd 1978, 152). Within the category of unhealthy waters are:

> stagnant waters from marshes and lakes [which] will necessarily be warm, thick and of an unpleasant smell in summer. Because such water is still and fed by rains, it is evaporated by the hot sun. Thus it is coloured, harmful and productive of biliousness. In winter it will be cold, icy and muddied by melting snow and ice. This makes it productive of phlegm and hoarseness (Lloyd 1978, 152).

The writer goes on to enumerate the features of the physiology and physiognomy of swamp dwellers.

The Loeb Classical Library translation of this essay also elaborates the description of the waters in the following miasmatic terms:

> turbid, stagnant waters in marshes and swamps are hot, thick and evil-smelling in summer because of their stagnation and failure to flow [...]

> Waters that issue level on the face of the earth and such as are marshy, standing and stagnant must in summer be hot, thick and stinking, because there is no outflow. (cited in Mattock and Lyons 1969, 48).

Not only does the turbidity of swamp waters mean that they mediate between solidity and liquidity (and so violate these categories), but their stagnancy and hotness violates the qualities of fluidity and coolness assigned to waters.

Waters, for the Hippocratic writer, should flow and be cool; waters which do not flow and are not cool violate the order of things. Hot, stagnant and turbid waters are also evil-smelling, whereas cool, flowing and clear waters are pleasant-smelling. Across a range of categories and qualities in the Hippocratic taxonomy of waters, wetlands are ruled out of court. On top of all this, and because of all this, they are bad for bodily health. For what reasons, the writer does not say, but there is a certain obviousness that this degree of category- and quality-mediation and violation is such a disruption to the order of things that it must be detrimental to health.

4.3 The Humours

For Aristotle the arrangement of the elementary qualities and their proper couples were associated with physical conditions and with mental humours or temperament. In his treatise 'De Problemata' in discussing 'problems connected with the effect of locality on temperament,' Aristotle goes on to answer with another question the question he poses:

> why is it that those who live in airy regions grow old slowly, but those who inhabit hollow and marshy districts age quickly? Is it because old age is a process of putrefaction, and that which is at rest putrefies, but that which is in motion is either quite free from, or at any rate less liable to, putrefaction, as we see in water? In lofty regions, therefore, owing to the free access of the breezes, the air is in motion, but in hollow districts it stagnates. Furthermore, in the former, owing to its movement, the air is always pure and constantly renewed, but in marshy districts it is stagnant. (Aristotle 1927, §909b)

Not only is the water stagnant in marshy districts, but the air as well. No wonder such districts are bad for you both physically and mentally!

Despite the differences between his work and that of his teacher, he reproduces, like a good pupil, Plato's denigration of 'hollow places' as Socrates undertook in the *Phaedo*:

> there are many hollow places all around the earth, places of every shape and size, into which the water and mist and air have collected. But the earth itself is as pure as the starry heaven in which it lies, and which is called Ether by most of our authorities. The water, mist, and air are the dregs of Ether, and they are continually draining into the hollow places of the earth. We do not realise that we are living in its hollows, but assume that we are living on the earth's surface [...] this earth and its stones and all the regions in which we live are marred and corroded by the brine, and there is no vegetation worth mentioning, and scarcely any degree of perfect formation, but only caverns and sand and measureless mud, and tracts of slime wherever there is earth as well; and nothing is in the least worthy to be judged beautiful by our standards. But the things above excel those of our world to a degree far greater still. (Plato 1954, §§109A–110D)

Socrates valorizes the extraterrestrial and the transcendental over the immanent—the intraterrestrial and the aquaterrestrial.

In the western theory of the elements, the four elements combine to form or are associated with psychosomatic states or 'humours' as developed by the Elizabethans and as formulated by E. M. W. Tillyard (1963, 76). These states can be tabulated in terms of the mixing of the four elementary qualities into the four legitimate elements, in turn into four more modern qualitative elements and finally into four proper substances and humours (Table 4.1):

This schema becomes problematic when considering the association between melancholia and wetlands as in the terms of this schema they

Table 4.1 The elements and the humours

	Hot	*Cold*
Dry	Fire	Earth
	Heat	Solid
	Choler/ic	Melancholy/ic
'Wet'	Air	Water
	Gaseous	Liquid/fluid
	Blood/sanguinous	Phlegm/atic

should not strictly be associated. Or more precisely, wetlands as between, or combining, earth and water should be seen as equally phlegmatic as they should be seen as melancholic. As a result, this combination of elements would produce a phlegmatic melancholy, an aberrant humour.

4.4 Monsters and the Monstrous

The wetland is also the home of monsters. In the pre-Socratic cosmogony, at least in Christian Athenagoras' second-century account, 'water was the origin for the totality of things [...] and from water slime was established, and from both of them was generated a living creature, a snake with a lion's head growing on to it,' visiting many of its wetlands and researching and writing about them[1] the king of anti-beasts, or monsters (Kirk et al. 1983, 25). The sixth labour of the heroic Heracles (Hercules) in Greek mythology is to kill the Stymphalian birds, 'the countless brazen-beaked, brazen-clawed, brazen-winged, man-eating birds, sacred to Ares' which 'had flocked to the Stymphalian Marsh' (Graves 1955, 119). They are waterbirds and war-birds—wa(te)rbirds for short! From the marsh, the wa(te)rbirds occasionally took to the air in 'great flocks to kill men and beasts by discharging a shower of brazen feathers and at the same time muting a poisonous excrement which blighted the crops' (Graves 1955, 119). No doubt this poisonous excrement is produced by the birds because they live in and feed greedily on the ostensibly decaying and rotting life of the marsh. In short, they are coprophagic.

The birds represent the antithesis of, or perhaps more precisely an offence (in at least two senses) towards, agriculture. In response, even resistance, to this myth my reading is a defence of the marsh and its birds. The marsh and the martial are allied in the myth against agriculture and men (andros). The ostensibly martial marsh-birds are figured as misandropic in this foundational myth of patriarchal western culture which displaces the blame onto the victim for the war men have fought against the marsh no doubt from before heroic Heracles' time. This martial anti-marsh campaign culminates in fascist Mussolini's boast that the war against

[1] Similarly the pre-Socratic 'Thales said that everything originated from water' (Cicero 1972, 80). Yet according to Aristotle for Thales everything 'not only originates from but returns to water which is thus both eternal (and so divine) and the stuff of which everything consists' (Hammond and Scullard 1970, 1050). See also Aristotle, *Metaphysics*, §983b20.

the Pontine Marshes was the kind of war he liked to fight, and win, at least in the short term.

Instead of men (and agriculture) being shown as antagonistic to the marsh, the marsh is made out to be monstrously martial towards men (and agriculture). A classic case of blaming the victim! The martial, men and agriculture (and the matrix) have been allied against the marsh, the maternal and marsh-birds in mainstream western culture. Marsh and agriculture are opposed but they can (or should) co-exist whereas the Stymphalian birds are made to represent a martial/marsh-al oral and anal sadism which would destroy the crops of the latter and vanquish the former over the latter. The oral and anal sadism which men have exercised against the marsh by eating it up, grazing their animals on it (consuming its resources), administering an enema to it (draining it) and crapping in it ('sanitary land-filling' it) is here displaced onto anthropophagic, or more precisely androphagic, coprophagic and coprophilic waterbirds. As they feed on shit and shit shit, their shit is not manure, but poison.

Heracles is initially unable to drive the birds away with his arrows. Typically, even archetypically, he finds that 'the marsh seemed neither solid enough to support a man walking, nor liquid enough for the use of a boat' (Graves 1955, 119). The wetland for Heracles, and for hundreds of generations of heroic westerners since him, is neither solid land nor deep water but mediates (between) the two. The wetland is centrally construed in mainstream western culture as a problem of transportation as it is for Allnutt in C. S. Forester's *African Queen* whose narrator remarks of the Swamp of the Bora that for Allnutt and Rose 'no place could be beautiful that presented navigational difficulties,' especially for the hero on a mission or a quest or with a job to do which requires him to traverse the place with some end or goal in view or mind (Forester 1956, 135). Yet rather than a hero on a single-minded mission to conquer the marsh by crossing it (though in one version of the myth Heracles drained the marsh (Graves 1955, 120)), Heracles' labour is to kill the horrific wa(te)rfoul.

This 'labour' 'glorifies' Heracles, according to Robert Graves, as:

> the healer who expels fever demons, identified with marsh-birds. The helmeted birds shown on Stymphalian coins are spoon-bills, cousins to the cranes which appear in English medieval carvings as sucking the breath of sick men. They are, in fact, bird-legged Sirens, personifications of fever [...] The spoon-bill is closely related to the ibis, another marsh-bird, sacred to the god Thoth, inventor of writing (Graves 1955, 120–121).

The Stymphalian birds personify, or more precisely 'femininify' as they were Sirens and women according to some accounts (Graves 1955, 120), the miasmatic diseases to which the marsh gives rise. How spoonbills, a benign species of waterbird which feeds on crustaceans and small fish by sweeping its spoon-shaped bill from side to side under the water, came to femininify disease and to be blamed for the war men have fought against the marsh is not only a mystery, but also a patent injustice (Kingsford 1991, 70).

A limited explanation of this mystery and injustice is offered by Bachelard (1983, 101) when he suggests that the marsh in this myth becomes what he calls:

> 'stymphalized.' It becomes the black swamp where live monstrous birds, the Stymphalides, 'nurslings of Ares, which shoot their feathers like arrows, which ravage and soil the fruits of the earth, which feed on human flesh.' This stymphalization [...] corresponds to a particular feature of melancholy imagination.

I suppose only a deeply melancholic, even manic depressive, imagination could see a spoonbill as a bird which wars on humans, destroys their crops by excreting on them and eats their flesh, as, in short, both anally and orally sadistic! As a result of such a diet, it is hardly surprising then that they 'mute' poisonous excrement! Such excrement is thus the result of their diet of human crops and flesh rather than the product of the marsh in which they live.

Marsh monsters are the product not only of a melancholic imagination but also, and in turn, of a miasmatic theory of monstrosity. Not only does the marsh 'cause' physical and 'mental' illness due to the vapours which rise from it, but also the miasma give 'birth' to monsters. Wetlands are dangerous to life from disease and can lead to death. They are even the land of the dead. Hades, the land of the dead, is a wetland for Homer. In *The Iliad* 'misty Tartaros' is 'the deepest gulf below earth' and in Hesiod's *Theogony* it is 'the grievous, dank limits which even the gods detest' (cited by Kirk et al. 1983, 9 and 40). Similarly in *The Odyssey*:

> [at ...] the limit [...]of the deep-running Ocean
> There lie the community and city of Kimmerian people,
> hidden in fog and cloud, nor does Helios, the radiant sun, ever break through the dark to illuminate them with his shining,

> neither when he climbs up into the starry heaven
> nor when he wheels to return again from heaven to earth,
> but always a glum night is spread over wretched mortals. (Homer 1965, XI 12–19, 168).

The ancient Romans took on and developed this view as we will see in the following chapter of the present volume.

4.5 Melancholy

Even though the melancholic has traditionally been associated with the cold and dry, a visit to the cold and wet/dry marshes by the drylander can induce melancholy, whereas the indigenous wetlander by temperament is usually phlegmatic, but tends towards the melancholic. By temperament, the wetlander at home in the wetland is phlegmatic (associated with the watery), whereas the drylander visiting the wetland tends to be infected, or more precisely affected, by the melancholic (the earthy, though more precisely the earthy in combination with the watery).

One of the 'causes of melancholy' for Robert Burton writing in the early seventeenth century in his *The Anatomy of Melancholy* was 'standing waters.' He went on to elaborate in Hippocratic terms that:

> standing waters, thick and ill-coloured, such as come forth of pools and moat where hemp is steeped or slimy fishes live, are most unwholesome, putrefied, and full of mites, creepers, slimy, muddy, unclean, corrupt, impure, by reason of the sun's heat and still standing; they cause foul distemperatures in the body and mind of man, are unfit to make drink of, to dress meat with, or to be used inwardly or outwardly [...] So that they that use filthy, standing, ill-coloured, thick, muddy water, must needs have muddy, ill-coloured, impure and infirm bodies. And because the body works upon the mind, they shall have grosser understandings, dull, foggy, melancholy spirits, and be readily subject to all manner of infirmities. (Burton 1932, 224–225)

Although Burton enunciates a psychosomatic theory of the body and mind (after all, he is pre-Descartes and Cartesian dualism), he subscribes to a simplistic analogy, or even correspondence theory, based primarily on colour, between the natural environment and the human body. And although he does not subscribe to the miasmatic theory of physical illness, it is the standing waters themselves, rather than the vapours which rise from them, which are to blame for illness, both 'mental' and physical.

Not only can standing waters cause melancholy, but also the cold and dry air which rises from them. When Burton discusses 'bad air' the worst air of all out of hot air, cold air, and cold and dry air is the last which he describes as:

> a thick, cloudy, misty, foggy air, or such as come from fens, moorish grounds, lakes, muckhills, draughts, sinks, where any carcasses or carrion lies, or from whence any stinking fulsome smell comes: [...] such air is unwholesome, and engenders melancholy, plagues and what not (Burton 1932, 239).

It is on such an 'anatomy' or miasmatic theory of melancholy and malaria that the Sanitary Movement of two centuries later was founded (as we will see later in the present chapter).

Yet such air is surely not dry but wet. Under the traditional western theory of the elements and humours as formulated systematically by the Elizabethans, the wetland did not fit easily into the schema as we have already seen. In fact, melancholy was strictly associated with the dry upland of the agricultural country and the modern city, and even with capitalism as Eagleton (1986, 41) argues that 'melancholia is an appropriate neurosis for a profit-based society' and with the sublime as Kant (1960, 47; see also 1952, 121) postulates that 'the feeling of the sublime is 'sometimes accompanied with a certain dread or melancholy.'

In response to the modern melancholic city sublime, there was a massive attempt to displace blame onto the wetland. The Elizabethan 'world picture' as enunciated by E. M. W. Tillyard was not flattering to the modern melancholic and sublimated city-dwelling drylanders; it was more flattering to the pre-modern phlegmatic and 'slimated' fen-dwelling wetlanders. As wetlands did not 'fit the picture,' rather than changing the picture, the wetland had to be changed, or more precisely in Burton's case the dwellers in wetlands, including in cities built on or near wetlands (invariably 'foreign'), denigrated.

After enumerating a list of places where the air is bad, Burton (1932, 240) concludes by asking 'how can they be excused that have a delicious seat, a pleasant air, and all that nature can afford, and yet through their own nastiness and sluttishness, immund and sordid manner of life, suffer their air to putrefy, and themselves to be choked up?' The wetland, its standing waters, its bad (but wet) airs and its filthy dwellers, either indigenous wetlanders or city-in-wetland dwellers, is the cause of melancholy displaced from the cold and dry uplands.[2]

4.6 Drain the Fens

This denigration of wetlands, especially in relation to health, was exported to, or imported into, the United Kingdom and was taken up by its writers in relation to the Fens of England. In Shakespeare's *King Lear*, Lear's ultimate repudiation of Cordelia is couched in the Hippocratic and miasmatic terms of 'Infect her beauty,/ You fen-sucked fogs.' In Shakespeare's *The Tempest*, Caliban curses Prospero in similar terms: 'All the infections that the sun sucks up/ From bogs, fens, flats, on Prosper fall, and make him/ By inchmeal a disease.' The Fens are employed in both instances as a source of insult against the health and beauty or life of someone who has been repudiated or is reviled.

The Fens are equated with disease, and its ravages, because of its bad airs. In John Carey's *New and Correct English Atlas* of 1787 the entry for Cambridgeshire remarked that 'in the Isle [of Ely in the Fens] the air is damp, foul, and unwholesome,' and for Norfolk that 'the marshy and watery places are aguish and unwholesome' (cited by Darby 1956, 176 n2). William Camden in his *Brittania* refers to 'the unwholesome aire of the Fennes' (cited by Darby 1956, 24). Yet it is not only the airs which are bad in the Fens. The Fens in fact mix the elements of earth, air and water:

> The moory soil, the wat'ry atmosphere,
> With damp, unhealthy moisture chills the air.
> Thick, stinking fogs, and noxious vapours fall,
> Agues and coughs are epidemical. (cited by Darby 1956, 176).

This mixture produces bad health according to William Dugdale:

> what expectation of health can there be to the bodies of men, where there is no element good? The air being for the most part cloudy, gross and full of rotten hairs [*sic*]; the water putrid and muddy, yea full of loathsome vermin, the earth spongy and boggy. (cited by Darby 1956, 58)

[2] For further discussion of melancholy and its association with wetlands, including in the twentieth century with such varied writers as Vera Brittain and E. M. Cioran, see Giblett (2021, Chap. 3).

As for the Hippocratic writer, wetlands violate the normative qualities of the elements: air should be clear and dry; water, clear and flowing; the earth, dry and solid. Why, as with the previous writers, these violations, these out of court mixings, should lead to ill-health is not made clear. Only a certain obviousness prevails that as they are such a disruption to the order of things across a number of categories, they must bode ill literally.

There were no shortage of writers who bemoaned the unhealthiness of the Fens and other English wetlands in accordance with the prevailing miasmatic theory and instances could be enumerated almost *ad infinitum*. A derelict bog was described in 1603 as having 'an atmosphere pregnant with pestilence and death' (cited by Coles 1989, 153). Over 200 years later Joseph Glynn referred to the Fens as 'the swamp or marsh, exhaling malaria, disease and death' (cited by Darby 1983, 176; see also Darby 1974, 9). If most writers were agreed about the cause of the unhealthiness, they were also agreed about the cure—drain the Fens. The miasmatic theory of disease produced an anti-ecology which destroyed habitats and wetlandscape in a bid ostensibly to improve health. Drainage of the Fens began in the Middle Ages and got going in earnest in the seventeenth century, especially precipitated by the 'Act for the Draining the Great Level Fens.' Fen drainage was to be carried out 'to the great advantage and strengthening of the nation' (cited by Darby 1956, 68). The colonizing and nationalistic aims of drainage were left in no doubt.

By the nineteenth century, despite the persistence of some fears, health improved. The eminent British historical geographer, H. C. Darby (1956, 180–182), who published three books on the Fens, remarked that:

> despite statistics of longevity, many people [in the nineteenth century] were still 'fearful of entering the fens of Cambridgeshire, lest the Marsh Miasma should shorten their lives.' There was considerable discussion as to what actually caused the ague-humid atmosphere, decaying vegetable matter, or bad drinking water. But whatever the cause might have been, it was certain that the complaint was disappearing. This improvement was generally ascribed to the better drainage.

So why enquire further into the cause of the ague when drainage was lessening the incidence? Why not keep draining? It is only in the twentieth century since the vector of the ague has been established that anyone has bothered to enquire into drainage and whether it was really necessary. In Graham Swift's *Waterland* his first-person narrator Crick pronounces 'the problem of the Fens has always been the problem of drainage,' but then wonders, 'is it desirable, in the first place, that land should be reclaimed?' (Swift 1984, 8).

But why not keep draining when it had such demonstrably positive effects besides the improvement of health? This was certainly the case for John Dyer whose poem 'The Fleece' was first published in 1757:

> Yet much may be performed to check the force
> Of Nature's rigour: the high heath, by trees
> Warm sheltered, may despise the rage of storms:
> Moors, bogs, and weeping fens, may learn to smile,
> And leave in dykes their soon-forgotten tears.
> Labor and art will every aim achieve
> Of noble bosoms. Bedford Level, erst
> A dreary pathless waste, [but then…]
> Russel, arose, who drained the rushy fen,
> Confined the waves, bid groves and gardens bloom,
> And through this new creation led the Ouse,
> And gentle Camus, silver-winding streams:
> Godlike beneficence: from chaos drear,
> To raise the garden and the shady grove. (cited by Darby 1956, 162)

Drainage could not only manage a recalcitrant nature and reverse the effects of the Fall; it could also produce a new, unfallen creation reversing history and returning the earth to the Garden of Eden as well as producing agricultural and moral 'improvements.' Such was the conclusion to which Arthur Young came in 1799 about 'the improvements' following embanking and enclosing in the Fens: '"fens of water, mud, wild fowl, frogs and agues, have been converted to rich pasture and arable" […] The result […] was "Health improved, morals corrected and the community enriched"' (cited by Darby 1983, 142). Young enumerates and equates what he sees as the worthless, useless, and the indigenous, if not downright harmful elements of the Fens, against which he poses monoculture and bourgeois culture. It is as if the drainage and discipline of the wetlandscape had a salutary and necessary connection with morality, the former producing the disciplining and draining of the unruly recalcitrance of indigenous populations, the latter restraining evil and inducing good morals.

Yet all writing about the Fens did not exclusively praise drainage. Probably the most passionate, articulate and ambivalent writer about the

Fens was Charles Kingsley. One of his *Prose Idylls* is devoted to the Fens and provides a highly ambivalent account of their drainage in which the discourses of the rationalist and the naturalist, the pastoral and the pastoralist contend against each other:

> reason and conscience tell us, that it is right and good that the Great Fen should have become, instead of a waste and howling wilderness, a garden of the Lord [...] And yet the fancy may linger, without blame, over the shining mere, the golden reed-beds, the countless water-fowl, the strange and gaudy insects, the wild nature, the mystery, the majesty [...] which haunted the deep fens for many a hundred years. (Kingsley 1884, 95)

Unlike his predecessor in Arthur Young who was happy to see everything and anything indigenous to the Fens transformed by drainage into monoculture, Kingsley both praises the benefits of 'progress' brought by drainage and bemoans simultaneously the losses inflicted on the Fens. Kingsley seems to want to have his cake of the old, pre-drainage Fens, and to eat his cake of the new, drained Fens, too. Or perhaps more precisely, he seems to want the good of the old Fens, such as the native flora and fauna, and not the bad, such as harshness and sickness, and to want the good of the new Fens, and not the bad.

But as the bad of the new Fens is tantamount to the loss of the old Fens, Kingsley is stuck in a bind, though for him the advantages of drainage ultimately outweigh the disadvantages:

> ah, well, at least we shall have wheat and mutton instead [of the native flora and fauna], and no more typhus and ague; and, it is to be hoped, no more brandy-drinking and opium-eating; and children will live and not die. For it was a hard place to live in, the old Fen (Kingsley 1884, 97).

In the Fens opium was known as 'the antidote to the effect of the noxious vapours' (cited by Dobson 1980, 370). But prosaic wheat and common mutton, even the eradication through drainage of what Kingsley calls 'the cold and poisonous fogs' (Kingsley 1884, 102) rising from the Fens, seem a poor swap, an inequitable exchange, for what he also calls poetically 'the shining meres, the golden reed-beds, the countless water-fowl, the strange and gaudy insects, the wild nature, the mystery, the majesty' (Kingsley 1884, 95; see also Kingsley 1877, 12–13). Similar sentiments were expressed at about the same time by Gerard Manley Hopkins in his poem 'Iversnaid':

> What would the world be, once bereft
> Of wet and of wildness? Let them be left,
> O let them be left, wildness and wet;
> Long live the weeds and the wilderness yet. (Hopkins 1953, 51).

This was the minority view in England, though it is one that was also propounded at about the same time in the United States by Sidney Lanier and Henry David Thoreau (see Giblett 1996, Chap. 10).

For Kingsley, however, unlike Hopkins, the bourgeois virtues of cleanliness, healthiness, godliness and temperance, though, outweigh any nostalgia for the lost wet and wild Fens:

> it is all gone; and it was good and right that it should go when it had done its work, and that the civilization of the fen should be taken up and carried out [...] till 'out of slough and bogs accursed, he made a garden of pleasure' (Kingsley 1884, 123).

The fen is 'a waste and howling wildernes' to be transformed into a pastoral/ist Eden, a barbarous place to be civilized by drainage, but as Thoreau (1988, 300) said, 'a howling wilderness does not howl: it is the imagination of the traveller that does the howling.'

Kingsley argues that the Fens constituted something of a first in unifying populations against nature as a common threat and a test-case of 'man [with large-scale industrial technology on 'his' side] versus ['the brute powers' (Kingsley 1884, 129) of] nature':

> it was in the fens, perhaps, that the necessity of combined effort for fighting the brute powers of nature first awakened public spirit and associate labor, and the sense of a common interest between men of different countries and races. But the progress was very slow; and the first civilisers of the fen were men who had nothing less in their minds than to conquer nature (Kingsley 1884, 117–118).

The drainage of the fens also gave for Darby (1956, 28) the phrase "man [*sic*] and his conquest of nature" its 'full meaning.' At least in the short term, but as Rachel Carson (1994, 86 and 188) wrote in *Silent*

Spring, the so-called conquest of nature is conceived in arrogance because 'man is part of nature.'

The conquest of wetlands by drainage was both militarist invasion and imperialist colonization. These were no mere metaphors as Swift's *Waterland* shows by its references to the 'endless and stationary war against water' (Swift 1984, 289), and to waging 'war on water, mud' (1984, 60). The war against wetlands was also no mere metaphor for the fascist Benito Mussolini in the 1930s with his scheme for draining the Pontine Marshes. These marshes have been described in a typical reductive gesture as 'nothing but a malaria-infested marsh' (Harris 1957, 311). During his Italian journey Goethe (1970, 180–181) also warned about their 'dangerous miasma,' though he thought they 'do not actually look as dreary as people in Rome usually describe them.' In Goethe's *Faust,* the dying speech of its eponymous hero endorses the modern and Faustian project of discipline and drain:

> A marshland flanks the mountain-side,
> Infecting all that we have gained;
> Our gain would reach its greatest pride
> If all this noisome bog were drained.
> I work that millions may possess this space,
> If not secure, a free and active race.
> Here man and beast, in green and fertile fields,
> Will know the joys that new-won region yields,
> Will settle on the firm slopes of a hill
> Raised by a bold and zealous people's skill.
> A paradise our closed-in land provides,
> Though to its margin rage the blustering tides;
> When they eat through, in fierce devouring flood,
> All swiftly join to make the damage good.
> Ay, in this thought I pledge my faith unswerving,
> Here wisdom speaks its final word and true,
> None is of freedom or life deserving
> Unless he daily conquers it anew. (Goethe 1969, 269)

The freedom and life of enclosed lands have to be conquered anew each day by conquering the oral sadistic flood which threatens to break the bulwarks erected against it. Paradise is enclosed land, recalling that Eden is a garden. The enclosed lands are located on the slopes of the hill (see Giblett 1996, Fig. 1, Chap. 2). They are thereby placed above, and right away from, the horrifying and threatening hell of the slimy wetlands with their bad airs and miasma rising from their black waters. The enclosed paradise is also positioned beneath, but not directly below, the terrifying heights of the sublime mountains and sublimated gases.

For Benito Mussolini the beneficial and 'peaceful' task of reclaiming the 'never-ending fen' of the Pontine Marshes was in his own words 'the kind of war which we prefer' (cited by Pini 1939, 180), perhaps because it was so easy to win, at least in the short term. According to Rachelle Mussolini (1959, 88), Benito's wife, to win 'the battle against the swamps,' especially the Pontine Marshes, was one of his five main objectives. Benito achieved this objective and later boasted that his two main achievements were that he had made the Italian trains run on time and that he had drained the Pontine Marshes. Italian fascist drainage of the Pontine Marshes merely extended and heightened processes already at work in patriarchy, modernity and modernization at large, in particular through the industrial conquest of the Earth.

Drainage has been seen as 'a major manifestation of progress,' as Purseglove (1989, 28) puts it. For Crick in Swift's *Waterland* 'my humble model for progress is the reclamation of land' (Swift 1984, 291) with its 'curious password: Drainage' (1984, 65). Whether 'progress' is a good or bad thing is another matter. Perhaps it was a 'good thing' then, but a 'bad thing' now. In the novel *Soundings* by Liam Davison (1993, 97, 99, 103 and 172), a kind of Australian *Waterland*, drainage 'put men in control' (control of what? one wonders, though the intransitivity might have been all that was desired) because it was 'Science put to a practical use' which produced 'a vision of Progress. It showed how the land could be mastered.' At least in the short term, but not in the long term with floods and storm surges. The effects of drainage on the (wet)landscape were profound. Clarence Glacken (1967, 348) in his monumental history of nature and culture in western thought from ancient times to the end of the eighteenth century argues that 'one of the truly great landscape changes in modern times has been marsh, bog, marine and lacustrine drainage.' I would go even further and suggest that wetlands' drainage is the landscape change which quintessentially characterizes modernity and modernization with all its cultural, corporeal and psychological ramifications, conditions of possibility and widespread effects. The fundamental event of the modern age was the draining and reclaiming of wetlands; the fundamental event of the hypermodern age is wetlands re-reclaiming their drained places.

4.7 The Sanitary Movement

Fears about the association between stagnant waters and malarial disease were one of the bases of the nineteenth-century Sanitary Movement which began in Great Britain under the tutelage of Edwin Chadwick and quickly spread to the colonies, such as the Swan River Colony of south-western Australia (Bolton and Hunt 1978). According to Peter Stallybrass and Allon White (1986, 125), Chadwick's *Report on the Sanitary Condition of the Laboring Population of Great Britain* first published in 1842 was 'an instant best-seller, and more than 10,000 copies were distributed free.' As Chadwick was a former literary assistant to Jeremy Bentham (Illich 1986, 45), he seems to have wanted to practice the principles of utilitarianism on the city. The Sanitary Movement applied Bentham's Panopticon principle, initially a design for a prison, to the hitherto invisible underside of the city, whether it be working class areas or to sewerage, in order to render it, or at least its effects, visible to health inspectors. By reading the effects in accordance with the interpretive grids of the Sanitary Movement, the causes could be deduced and remedies prescribed. The Movement was also based on the utopian and educative impulse that good drainage meant good morals. It subscribed to the 'discipline and drain' school of thought. It was thus part of the general move in the nineteenth century to police, regulate and place under surveillance the life of populations (Foucault 1977).

The founding trope of the Sanitary Movement was the working class abode figured as tropical swamp. Southwood Smith, described as 'perhaps the most influential of the medical pioneers' of the Sanitary Movement, drew the analogy in his *Treatise on Fever* published in 1830:

> the room of a fever-patient, in a small and heated apartment in London, with no perflation of fresh air, is perfectly analogous to a stagnant pool in Ethiopia, full of the bodies of dead locusts. The poison generated in both cases is the same; the difference is merely in the degree of its potency. Nature with her burning sun, her stilled and pent-up wind, her stagnant and teeming marsh, manufactures plague on a large and fearful scale. (cited by Greenwood 1953, 502)

Heat is the agent which produces malignancy, as it is in C. S. Forester's *The African Queen* in which the sun blazes malignantly (Forester 1956, 111). As heat is the agent which transforms solid into gas (in sublimation) and solid into liquid (in slime), so is it also that which changes liquid into gas (in malaria or miasma). In Rider Haggard's *She* 'the mist was being sucked up by the sun' (Haggard 1991, 61). In the tropical swamp of the Bora of *African Queen* 'the heat was colossal' (Forester 1956, 106). Put this together with 'the stifling air' (Forester 1956, 106) and the result is 'the stifling heat' (107). But the result is not just any heat and certainly not dry heat. What results, in turn, is 'the dampness of the heat' (107), the heat acting on water to produce 'that inferno of sticky heat' (108) of 'the fever of the day' (116) of the tropical swamp.

The other within (the inner-city slum, the working class) is analogous to, or the same as for all intents and purposes, as the other without (the tropical swamp, 'the natives'). As 'nature' and 'the natives' were colonized by European imperialism with its medical and other technologies, including engineering, so were the tropical swamp and the inner-city slum. The working class, an 'other' within, was even seen by Chadwick (1965, 158) as, and reduced to, agricultural animals, another colonized other within: 'the precautions applicable to the sheep and cattle will be deemed equally applicable to the laboring population.' Although Friedrich Engels (1969, 31, 32; see also 34) in his 'Preface to the English Edition' of *The Condition of the Working Class in England* did not draw the same analogy as Southwood Smith, he used the same metaphor in describing the East End of London as 'an ever-spreading pool of stagnant misery and desolation.' By a neat, but not nice, historical irony the area of the city built on drained marshes reverts to the swamp of working class urban areas. Such is the persistence of the association between disease and the working class 'swamp' that when Anthony Wohl (1984, 286) introduced Southwood Smith's analogy he did so by describing 'the slum dwelling as potentially lethal as any disease-ridden tropical marsh.' Similarly Eric Hobsbawm (1985, 249 and 232) has luridly referred to what he calls 'the slimy pools of sweated and casual poverty.'

The Sanitary Movement was clearly based on the miasmatic theory of disease because for it, like for C. S. Forester later in *The African Queen*, 'smells generated disease' (Flinn, 'Introduction,' Chadwick 1965, 62). As the title of Forester's novel suggests, it is set in Africa and the smells referred to arise from a swamp, the Swamp of the Bora. First published in 1936 and set in 1914, so more than 20 years after the discovery of the vector of malaria, this 'classic first world war love story' and 'the all-time favourite' according to the Penguin cover blurb, perpetuates the miasmatic theory. It became the basis for the famous film of the same title starring Humphrey Bogart and Katherine Hepburn and directed by John Huston. It was the highest grossing film of the 1950s.

In the novel the swamp of the Bora is described as 'a pestilential fever swamp [...] steaming in a tropical heat, overgrown with dense vegetation, the home of very little animal life, and pestilent with insects,' though insects are animals (Forester 1956, 105–106). It may not have been the home of much charismatic macrofaunal life, but no doubt it was the home of a myriad, but banal microfaunal life. Part of 'the horror of the place' (Forester 1956, 137) consists in it being neither land nor water but between the two, as the horror of the slimy for Sartre resides in part in it being between solid and liquid, neither one nor the other. The swamp is between land and water as slime is between solid and liquid (as the sublime is between solid and gas) (see Giblett 1996, Fig. 1, Chap. 2).

The mixing of earth and water marks a transition from the dreariness associated with the mixture of land and water to outright horror, especially when this mixture invades the 'African Queen' 'herself':

> it was a nightmare time of filth and sludge and stench. Be as careful as they would, the all-pervading mud spread by degrees over everything in and upon the boat, upon themselves, everywhere, and with it came its sickening stench. It was a place of twilight, where everything had to be looked at twice to make sure what it was, so that, as every step might disturb a snake whose bite would be death, their flounderings in the mud were of necessity cautious. Worse than anything else it was a place of malaria (Forester 1956, 138).

Whilst Forester never explicitly states that the stench of the swamp caused the malaria, calling it 'a place of malaria' is not equivalent to suggesting it was a place of the mosquito that is the vector for the parasite that infects humans with malaria. Similarly although the swamps of Africa in Rider Haggard's *She* may look bad and be depressing, their worst feature is their smell which were thought to carry disease:

> undoubtedly the worst feature of the swamp was the awful smell of rotting vegetation that hung about it, which was at times positively overpowering, and the malarious exhalations that accompanied it, which we were of course obliged to breathe. (Haggard 1991, 60)

The association reproduced here between stench and disease was the foundation of the Sanitary Movement in nineteenth-century Britain. The reference to 'a dank smell of rotting vegetation' in *The African Queen* (Forester 1956, 129) could equally be found in the literature of the Sanitary Movement one hundred years before with its references to 'putrescent animal and vegetable matter.'

The Sanitary Movement subscribed to the view that for Chadwick 'all smell is disease' (cited by Stallybrass and White 1986, 139). For Chadwick and the Sanitary Movement, as Roy Porter (1994, vi), points out, 'stench was, in fact, disease.'[3] This conclusion must have been the result of and have entailed a massive repression and sublimation of the olfactory in which all smells were seen as bad, though generally it was only bad smells which were thought in the miasmatic theory to have caused disease. For the Sanitary Movement disease was thought to have been 'generated in the miasma given off by decaying organic matter' (Flinn, 'Introduction,' Chadwick 1965, 62). Miasma, as David Miller (1989, 190–192) has pointed out, are 'an infectious air said to emanate from decaying matter and swamps and marshes' and 'originally stood for any pollution or polluting agent' (Parker 1983, especially 1–17). Malaria and marsh miasmata were, in fact, interchangeable terms (Ferguson 1823, 273; Curtin 1964, 182–188). In 1830 Dr Edwin D. Faust claimed that 'malaria may be considered as an uncombined gas, as the vapour of a volatile solid or liquid, as animalcular, or as a gas combined with water' (cited by Jarcho 1970, 36). Malaria, like miasma, = air (or gas) + water.[4]

The cure for Chadwick (1965, 99) was simple, in a word, drainage, as well as garbage collection. Or, if wetlands are the cause of the disease, in the terms of one of Chadwick's authorities, it was a simple matter of 'suppress the marsh to expel the fever,' the sloganeering cure prescribed by the Sanitary Movement (Chadwick 1965, 161). Drainage produced the double benefit of salubrious air and productive soil. It could even be applied successfully to the age-old problem of the Isle of Ely in the Fens and Chadwick quotes approvingly John Marshall:

> it has been shown that the Isle of Ely was at one period in a desolate state, being frequently inundated by the upland waters, and destitute of the adequate means of drainage; the lower parts became a wilderness of stagnant pools, the exhalations from which loaded the air with pestiferous vapours

[3] Corbin (1994) discusses miasma, effluvia, marshes and swamps extensively but, surprisingly, makes no reference to malaria.

[4] For further discussion of the history of malaria see Covell (1967), Bruce-Chwatt (1976, 1988) and Harrison (1978).

> and fogs; now […], by the labor, industry, and spirit of the inhabitants, a forlorn waste has been converted into pleasant and fertile pastures, and they themselves have been rewarded by bounteous harvests. Drainage, embankments, engines, and enclosures have given stability to the soil […] as well as salubrity to the air. These very considerable improvements, though carried on at a great expense, have at last turned to a double account, both in reclaiming much ground and improving the rest, and in contributing to the healthiness of the inhabitants. Works of modern refinement have given a totally different face and character to this once neglected spot. (Chadwick 1965, 150–151; see also 153)

In other words, modernization has converted primeval wilderness and modern wasteland into pastoral productivity. Unlike Kingsley who bemoaned the loss of the old Fens, Marshall is completely gung-ho about drainage.

As for Arthur Young in the eighteenth century, so for the Sanitary Movement in the nineteenth, suppressing the marsh and draining the fen not only expelled fevers, improved health and produced fertile land for agriculture, but also reformed morals. Chadwick asked 'how much of rebellion, of moral depravity and of crime has its root in physical disorder and depravity?' (cited by Stallybrass and White 1986, 131). He went on to postulate that 'the fever nests and seats of physical depravity are also the seats of moral depravity, disorder and crime with which the police have most to do' (cited by Stallybrass and White 1986, 131). Unlike Southwood Smith who only drew a parallel between the spaces of the tropical swamp and working class abode, Chadwick goes further to draw an analogy between the physical and moral.

For Chadwick (1965, 205), not only were 'the powerful operation of depraved domestic habits [seen] as a predisposing cause to disease', but also 'the noxious physical agencies depress the health and bodily condition of the population and act as obstacles to education and to moral culture' (Chadwick 1965, 268). The moral could be corrected by correcting the physical. The way to a 'man's' heart was not only through his stomach, but also through his nose. In order to foster moral culture, it was simply a matter of draining stagnant pools: 'the cases of moral improvement of a population, by cleansing, draining and the improvement of the internal and external conditions of the dwelling' (Chadwick 1965, 200). In this way 'the effects of the noxious physical agencies on the moral condition of the population' (Chadwick 1965, 204) could be ameliorated.

By a poignant historical irony, or perhaps poetic justice, the engineering which destroyed wetlands through drainage, and so had some initial ameliorating effect on the incidence of malaria, has contributed also to the re-introduction of malaria through the building of dams and the creation of artificial lakes, ideal breeding grounds for mosquitoes, a conclusion to which at least one writer in the early nineteenth century had come (Collins 1829, 30–31) and about which one late twentieth-century writer warned (Waddy 1975, 8). The position of engineering in regard to wetlands encapsulates the contradictory nature of modernization and modernity: on the one hand, the power of engineering to alter the (wet)landscape, to drain wetlands and so to ameliorate malaria; and on the other to build dams, to create artificial lakes and so to re-introduce malaria. This paradox also highlights a shift in the social status of the engineer from modern hero to postmodern Faust who has sold his (and usually it is his) soul to the Devil of hyper-capitalism in return for power and money at the expense of, amongst other places, the wetlands and their indigenous populations of people, fauna and flora, and the mystery and majesty of wet wilderness.

References

Aristotle. 1927. Problemata. In *The Works, Volume VII*, ed. W.D. Ross. Trans. E.S. Forster. Oxford: Clarendon Press.

———. 1930. De Generatione et Corruptione. In *The Works, Volume II*, ed. W.D. Ross. Trans. H.H. Joachim. Oxford: Clarendon Press.

Bachelard, Gaston. 1983. *Water and Dreams: An Essay on the Imagination of Matter*. Trans. E.R. Farrell. Dallas, TX: The Pegasus Foundation/The Dallas Institute of Humanities and Culture.

Bolton, Geoffrey, and Su-Jane Hunt. 1978. Cleansing the Dunghill: Water Supply and Sanitation in Perth 1878–1912. *Studies in Western Australian History* 2: 1–17.

Bruce-Chwatt, L.J. 1976. Ague as Malaria. An Essay on the History of Two Medical Terms. *The Journal of Tropical Medicine and Hygiene* 79: 168–176.

———. 1988. History of Malaria from Prehistory to Eradication. In *Malaria: Principles and Practice of Malariology*, ed. Walther H. Wernsdorfer and Ian McGregor, 1–59. Edinburgh: Churchill Livingstone.

Burton, Robert. 1932. In *The Anatomy of Melancholy*, ed. Holbrook Jackson. London: Dent. First published 1621.

Carson, Rachel. 1994. *Silent Spring*. Boston, MA: Houghton Mifflin. First published 1962.

Chadwick, Edwin. 1965. In *Report on the Sanitary Condition of the Laboring Population of Great Britain*, ed. M.W. Flinn. Edinburgh: Edinburgh University Press. First published 1842.

Cicero. 1972. *The Nature of the Gods*. Trans. H. McGregor. London: Penguin.

Coles, Bryony and John. 1989. *People of the Wetlands: Bogs, Bodies and Lake-Dwellers*. London: Thames & Hudson.

Collins, S.H. 1829. *The Emigrants Guide to the United States of America*. Hull: Joseph Noble.

Corbin, Alain. 1994. *The Foul and the Fragrant: Odour and the Social Imagination*. London: Picador.

Covell, Gordon. 1967. The Story of Malaria. *The Journal of Tropical Medicine and Hygiene* 70: 281–285.

Curtin, Philip D. 1964. *The Image of Africa: British Ideas and Action, 1780–1850*. Madison, WI: University of Wisconsin Press.

Darby, H.C. 1956. *The Draining of the Fens*. 2nd ed. Cambridge: Cambridge University Press.

———. 1974. *The Medieval Fenland*. 2nd ed. Newton Abbot: David and Charles.

———. 1983. *The Changing Fenland*. Cambridge: Cambridge University Press.

Davison, Liam. 1993. *Soundings*. St Lucia: University of Queensland Press.

Dobson, Mary. 1980. 'Marsh Fever' – The Geography of Malaria in England. *Journal of Historical Geography* 6 (4): 357–389.

Eagleton, Terry. 1986. *William Shakespeare*. Oxford: Basil Blackwell.

Engels, Fredrich. 1969. *The Condition of the Working Class in England*. London: Grafton. First published 1882.

Ferguson, William. 1823. On the Nature and History of the Marsh Poison. *Transactions of the Royal Society of Edinburgh* 9 (2): 273–298.

Forester, C.S. 1956. *The African Queen*. Harmondsworth: Penguin. First published 1935.

Foucault, Michel. 1977. *Discipline and Punish: The Birth of the Prison*. Trans. A. Sheridan. Harmondsworth: Penguin. First published in French in 1975.

Giblett, Rod. 1996. *Postmodern Wetlands: Culture, History, Ecology*. Edinburgh: Edinburgh University Press.

———. 2021. *Wetlands and Western Cultures: Denigration to Conservation*. Lanham, MD: Lexington Books.

Glacken, Clarence. 1967. *Traces on the Rhodian Shore: Nature and Culture in Western Thought from Ancient Times to the End of the Eighteenth Century*. Berkeley: University of California Press.

Goethe, Johann Wolfgang von. 1969. *Faust*, Part Two. Trans. P. Wayne. Harmondsworth: Penguin. First published in German in 1832.

———. 1970. *Italian Journey*. Translated by W. H. Auden and Elizabeth Mayer. Harmondsworth: Penguin.

Graves, Robert. 1955. *The Greek Myths, Volume II*. Harmondsworth: Penguin.

Greenwood, Major. 1953. Miasma and Contagion. In *Science, Medicine and History: Essays on the Evolution of Scientific Thought and Medical Practice Written in Honour of Charles Singer, Volume II*, ed. E. Ashworth Underwood, 501–507. London: Oxford University Press.

Haggard, H. Rider. 1991. *She*. Oxford: Oxford University Press. First published 1887.

Hammond, N.G.L., and H.H. Scullard, eds. 1970. *Oxford Classical Dictionary*. 2nd ed. Oxford: Clarendon Press.

Harris, L.E. 1957. Land Drainage and Reclamation. In *A History of Technology, Volume III*, ed. Charles Singer, E.J. Holmyard, A.R. Hall, and Trevor J. Williams, 300–323. Oxford: Clarendon Press.

Harrison, Gordon. 1978. *Mosquitoes, Malaria and Man: A History of the Hostilities Since 1880*. New York: E. P. Dutton.

Hobsbawm, E.J. 1985. *The Age of Capital 1848–1875*. London: Abacus.

Homer. 1965. *Odyssey*. Trans. R. Lattimore. New York: Harper and Row.

Hopkins, Gerard Manley. 1953. *Poems and Prose*. Harmondsworth: Penguin.

Illich, Ivan. 1986. *H_2O and the Waters of Forgetfulness*. London: Marion Boyars.

Jarcho, Saul. 1970. A Cartographic and Literary Study of the Word *Malaria*. *Journal of the History of Medicine* 25 (1): 31–39.

Kant, Immanuel. 1952. *The Critique of Judgement*. Trans. J.C. Meredith. Oxford: Clarendon Press.

———. 1960. *Observations on the Feeling of the Beautiful and Sublime*. Trans. J.T. Goldthwait. Berkeley: University of California Press.

Kingsford, Richard. 1991. *Australian Waterbirds: A Field Guide*. Kenthurst, N.S.W: Kangaroo Press.

Kingsley, Charles. 1877. *His Letters and Memories of His Life, Volume 1*, edited by his wife. London: Henry S. King.

———. 1884. The Fens. In *The Works, Volume XV, Prose Idylls*, 93–131. London: Macmillan.

Kirk, G.S., J.E. Raven, and M. Schofield, eds. 1983. *The Presocratic Philosophers: A Critical History with a Selection of Texts*. 2nd ed. Cambridge: Cambridge University Press.

Lloyd, G.E.R., ed. 1978. *Hippocratic Writings*. Trans. J. Chadwick, W. N. Mann, I.M. Lonie and E.T. Withington. Harmondsworth: Penguin.

Mattock, J.N. and M.C. Lyons, ed. and trans. 1969. *Hippocrates: On Endemic Diseases, Airs, Waters and Places*. Arabic Technical and Scientific Texts, 5. Cambridge: W. Heffer.

Merchant, Carolyn. 1980. *The Death of Nature: Women, Ecology and the Scientific Revolution*. San Francisco, CA: Harper & Row.

Miller, David. 1989. *Dark Eden: The Swamp in Nineteenth-Century American Culture*. Cambridge: Cambridge University Press.

Mitchell, John G. 1992. Our Disappearing Wetlands. *National Geographic* 182 (4): 2–45.

Mussolini, Rachelle. 1959. *My Life With Mussolini.* London: Robert Hale.

Parker, Robert. 1983. *Miasma: Pollution and Purification in Early Greek Religion.* Oxford: Clarendon Press.

Pini, Giorgio. 1939. *The Official Life of Benito Mussolini.* Trans. L. Villari. London: Hutchinson.

Plato. 1954. Phaedo. In *The Last Days of Socrates*, 93–185. Trans. H. Tredennick. Harmondsworth: Penguin.

Porter, Roy. 1994. Foreword: Alain Corbin. In *The Foul and the Fragrant: Odour and the Social Imagination*, v–vii. London: Picador.

Purseglove, Jeremy. 1989. *Taming the Flood: A History and Natural History of Rivers and Wetlands.* Oxford: Oxford University Press.

Ross, Sir David. 1995. *Aristotle.* 6th ed. London: Routledge.

Stallybrass, Peter, and Allon White. 1986. *The Politics and Poetics of Transgression.* London: Methuen.

Swift, Graham. 1984. *Waterland.* London: Picador.

Thoreau, Henry David. 1988. *The Maine Woods.* New York: Penguin. First published 1864.

Tillyard, E.M.W. 1963. *The Elizabethan World Picture.* Harmondsworth: Penguin. First published 1943.

Waddy, B.B. 1975. Mosquitoes, Malaria and Man. In *Man-Made Lakes and Human Health*, ed. N.F. Stanley and M.P. Alpers, 7–20. London: Academic Press.

Wohl, Anthony S. 1984. *Endangered Lives: Public Health in Victorian Britain.* London: Methuen.

CHAPTER 5

Paludal Romans and their Dispersion

Ancient Roman culture acknowledged the vital role of wetlands in the past in both the creation of life and the foundation of Rome. Yet that is where and when this acknowledgement begins and ends. Ancient Roman culture defined its trajectory from the past to the present and on into the future as emerging from the wetland and leaving it behind spatially with the development temporally of agriculture, pastoralism, towns, cities and empire. It defined and constituted all of the latter as distinct and apart from the paludal spatially, placially, temporally and culturally Ancient Roman culture was anti-paludal in the present and with is present places. It placed the paludal in the past, as the ante-place before the present.

In addition, the farm, town, city and empire are dry land on the surface of the earth, whereas the paludal is wet land below the surface in the depths. The former for ancient Roman culture is the place of managed and civilized life, whereas the latter is the place of monstrous and barbarous death. The farm, town, city and empire are *not* the wetland in a binarist opposition; the former are +1, or positive and presence, the wetland is 0, or lack and absence; the former fills the latter (often literally). In short, the farm, town, city and empire are culture, and the wetland (among other places and their peoples) is nature. The farm, town, city and empire versus the wetland encapsulates the culture-nature binary opposition, a fundamental strata of bedrock for mainstream western culture and its

R. Giblett, *Wetland Cultures*, Literatures, Cultures, and the Environment, https://doi.org/10.1007/978-3-031-57365-1_5

denigration of wetlands and other places of nature. The present chapter traces the history of anti-paludal Roman culture from ancient times through their dispersion into Medieval Italian culture (Dante Alighieri), English Renaissance and Reformation culture (John Milton), and beyond into the nineteenth and twentieth centuries, concluding with the presentation of some contemporary pro-paludal writing.

5.1 Anti-paludal Roman Culture

Ancient Roman culture gave wetlands a role in the past, in the beginning of life, in nature's watery womb. Ovid in his *Metamorphoses* repeats the myth of the flood and relates that:

> after that old moisture remaining from the flood had grown warm from the rays of the sun, the slime of the wet marshes (*paludes*) swelled with heat, and the fertile seeds of life, nourished in that life-giving soil, as in a mother's womb, grew and in time took on some special form. (Ovid 1984, I, 418–421, 30–33)

Marshes are slimy, fertile, maternal and womby places for the creation of life. This association was later to give rise to proto-fascist and fascist hostility to marshes and female bodies (as we saw in previous chapters). The association between marshes and heat was also used to give credence to the miasmatic theory of disease in both the tropics and the slums in the Sanitary Movement (as we saw in the previous chapter). Ancient Roman culture also gave wetlands a role in the past with the foundation myth of Rome (as we will see in the present chapter). Ancient Roman culture had a myth of triumphal, imperial progression out of the slimy, marshy and maternal past and places into the sublime and dry present and places, and for future. The trajectory of ancient Roman culture was sublimation upwards and onwards, out of slime and into the sublime.[1]

Longinus in the first century of the common era enunciated this upward and forward, spatial and temporal trajectory in his treatise 'On the Sublime.' The sublime for Longinus (1965) 'uplifts our souls' (107), fills one with 'a proud exaltation' (107), is associated with a sense of 'vaunting

[1] For further discussion of sublimation as a psychological process and philosophical category, see Giblett (1996, Chap. 2).

joy' (109) and is 'a noble emotion [that] forces its way to the surface in a gust of frenzy' (109). The sublime involves 'the production of grand ideas, perpetually impregnating them with a noble inspiration' (109). It impregnates one with heavenly power (119). Finally, the sublime carries one up close to the majestic mind of God (147). It carries 'one' up, up and away from the watery womb of the goddess. The sublime is an artificial womb that impregnates 'one' with heavenly power close to the divine majestic mind out of which 'one' gives birth to grand ideas as portrayed in the birth of Pallas Athene from the head of Zeus. The sublime is the obverse or secret of slime that gives birth to life out of nature's watery womb and earthly prowess. Slime as the secret of the sublime is captured brilliantly in Zoë Sofoulis's parenthetical portmanteau word 's(ub)lime' (Sofoulis 1988, 15, n86).[2]

The ancient Roman process of sublimation took place in both the city and the country, and in the country out of the paludal into the agricultural and pastoralist. Ancient Roman culture was agricultural and pastoralist; it was not paludicultural. It did not cultivate local, native (endemic) species of plants in wetlands for human and animal consumption. The likely candidate of rice as a wetland grain is ruled out of contention as the exception as it was introduced to Italy long after the fall of the Roman empire. The earliest documented cultivation of rice in Italy was in the wetlands of Tuscany, near Pisa, in 1468 CE, roughly a thousand years after the fall of the Roman empire (Tinarelli 2006). Rice has since played an important role in Italian agriculture with the cultivation of its own varieties, most famously Arborio, and in Italian cuisine, especially risotto, perhaps the ultimate comfort food, certainly up there with pasta and pizza.

Ancient Romans for Varro (1935, 424–425) 'called the same earth "mother" and "Ceres",' 'properly "the creator",' according to a footnote. They conflated and confounded creation, the growing of cereals and mother earth. They had no place for the great mother of the swamps. They defined agriculture and pastoralism against the *paludes* (marshes, meres). Agriculture and pastoralism were spatially and culturally antithetical to the paludal. Agriculture and pastoralism were markers of distinction—a value-laden distinction—from the paludal. Varro stated that '*Agri cultura*' is 'tilling of the ground' performed by oxen and the ploughman (Varro 1935, 162–163). By contrast, pastoralism is the grazing of herds of sheep and cattle on grasses growing in drylands and tended by the

[2] For further discussion of the s(ub)lime, see Giblett (1996, Chap. 2).

herdsman and the shepherd. Agriculture and pastoralism are located above ground, the dry surface of the earth, whereas the paludal is the domain of the below ground, the wet depths of the earth. The paludal is not good for agriculture and pastoralism, though 'a place which is swampy (*palustrem*)' is good for keeping flocks of ducks (Varro 1935, 486–487) in aquaorniculture, the cultivation of waterbirds for human consumption.

As the paludal is below ground it was only a simple step to regard it as the land of the dead. For Homer and Virgil the underworld is a swamp. Dante inherited this view from Virgil and Christianized it. Dante followed in Virgil's footsteps, not only on his journey from hell through purgatory to paradise, but also in writing his descriptions of these places, as did Milton in the seventeenth century. Rather than hell is a swamp, for later Anglophone writers in the nineteenth and twentieth centuries the swamp is hell. It is a place of death and destruction. For twenty-first century writers attuned to wetland ecology, such as Delia Owens in her recent novel, *Where the Crawdads Sing*, swamps are places of both life and death (as we will see later in the present chapter).

The ancient Romans inherited the malarial Greeks' views of the link between marshes and miasma. The ancient Romans also inherited the malarial Greeks' views of the link between marshes and monsters. When Lucretius (1994, 129) discussed the nature of the universe in the first century before the common era (BCE) he described the Stymphalian birds of Herculean myth as 'the foul birds that haunted the Stymphalian lake.' A later translation calls them 'dire fowl' (Lucretius 2007, 148). They are waterfowl of foul water (as we saw in the previous chapter). Lucretius also subscribed to the miasmatic theory of disease, though he did not associate miasma specifically with wetlands (Lucretius, 1994, 194–195; 2007, 232). Pliny the Elder did both, though, when he related how:

> doctors investigate which kind of waters are beneficial. They rightly condemn stagnant, sluggish waters, and regard running water as more beneficial because it is rendered fine and healthy by the agitation of the current [...] Wholesome waters should also be without taste or smell. (Pliny the Elder 1991, 274)[3]

Pliny the Elder (1942, 621) was following in the footsteps of Hippocrates who he called 'that prince of medicine.' Pliny also relates that

[3] For further discussion, see Borca (2000, 75).

Hippocrates was given the same honours as were given to the heroic Hercules, the killer of waterbirds (Pliny the Elder 1942, 589).

Unwholesome waters are stagnant, yet stagnant waters are also a refuge, especially in the myth of the founders of Rome and Roman culture about the twins Romulus and Remus. Rome, like many other cities before and since, was founded in a marsh, or at least it was in its foundational myth.[4] Borca (1997, 6) relates that Romulus and Remus were:

> exposed to a marsh next to the overflowing Tiber [...], probably the *Velabrum* which [...] assumed a marked infernal character [...] stagnant waters, indeed, keep the children from being abandoned to the real current of the river, save the divine twins from death, and thus let them accomplish the mission of establishing the culture [of Rome]. In the ambiguity of a wasteland [or wetland], whose edges are indistinct and which is placed midway between earth and water, the children's contact with nature and infernal powers becomes even more important. Border area, but not neatly bordered itself, the marsh is the safest refuge to those who subvert the old order and establish the new cosmos [or empire].

The marsh is a refuge, as it was for Octavian/Augustus (later Caesar) who hid for three days in a marsh (*palude*) during the Battle of Philippi (Pliny the Elder 1942, 604–605).[5] This marsh and other wetlands are sites of strategic and tactical importance in warfare as they are easy to defend and hard to attack.[6] The marsh is also a site of resistance against invaders, including by the Britons against the Romans, the Danes and the Normans (as we will see later in the present chapter).

The ancient Roman cultural trajectory upward and forward out of the wetland, onto the dry land of the farm and into the city is encapsulated in Cicero's idea of second nature. In the first century BCE, Cicero drew the distinction between the first nature of the natural world and the second nature of the human world of herding, tilling, mining, forging and building. Second nature works over first nature. For Cicero second nature involved cultivation and agriculture, settlements and cities, weaving and sewing, mining and metallurgy. Agri-culture is nature made over into second nature which occurs not only 'in the cultivation of the soil,' as Cicero

[4] For further discussion of cites founded in wetlands, see Borca (2000, 74) and Giblett (2016).

[5] For further discussion of wetlands as refuge, see Giblett (1996, Chap. 9).

[6] For further discussion of wetlands in warfare, see Giblett (1996, Chap. 9).

(1972, 184–185) put it, but also 'in the building of houses, in the weaving and sewing of our clothes and in the working of bronze and iron,' all the means by which 'we seek with our human hands to create a second nature in the natural world.' They are also the means by which 'we' to create a second culture in the natural world. Second nature (and culture) involve permanent dwellings, agriculture, mining of minerals and forging them into metals. It also involved the working over of the first nature of first peoples.

Cicero's distinction smacks of the doctrine of *terra nullius* applied most famously to Australia by Captain James Cook that denies the work of Aboriginal hunter-gatherers, nomads, engineers of aquaculture, graziers of pasturalism and farmers of paludiculture in working nature and producing first nature. By placing second nature *in* the natural world and by not making a distinction between nature and first nature, Cicero equates first peoples with nature, and so denies their work in shaping and using the land and erases them. The means by which 'we' create a human world *in* the natural world are also the means by which 'we' create second nature (and culture) *on* first nature and culture. Rather than the opposition between culture and nature, the crucial distinction for understanding the relationship between human beings and the earth, and achieving earthly mutuality between them, is that between the cultures of first, or worked, nature and the cultures of second, or worked-over, nature.[7] Relational views of nature, such as this of the cultures of natures, are now gaining ground in global conservation policy debates and scholarship. They provide one context for the present volume.

5.2 Hell as Swamp

In contrast to the other ancient Roman writers already mentioned, Virgil went one step further and focused on the infernal character of marshes. In the *Georgics*, he paints a picture of 'black Styx and the shades infernal' where 'with his tortuous coils, the mighty Snake glides forth, river-like' (Virgil 1999, I, 243–244, 117). The Styx, the river of the underworld, is a black serpent. In 'the land of the dead,' Dis is 'the grove that is murky with black terror' (1999, IV, 467–469, 253). From 'the lowest realms of Erebus,' or darkness itself or himself, 'came the insubstantial shades, the phantoms of those who lie in darkness' (1999, IV, 471–472, 253) while:

[7] For further discussion of the cultures of natures, see Giblett (2011, Chap. 1).

round them are the black ooze and unsightly reeds of Cocytus, the unlovely mere (*palus*) enchaining them with its sluggish water, and Styx holding them fast within his ninefold circles. (1999, IV, 478–480, 252–253)

Virgil transforms Cocytus, a river surrounding the underworld, into marsh. This is 'the very house of Death and deepest abysses of Hell' (1999, IV, 481–482).

In the *Aeneid* Virgil paints a similar picture of 'the waters of Tartarean Acheron,' one of the rivers of the underworld, where, 'thick with mire and of fathomless flood, a whirlpool seethes and belches into Cocytus all its sand' (1999, VI, 295–298, 553). This is 'the Stygian marsh (*Stygiamque paludem*)' (1999, VI, 323, 554–555; see also lines 369, 558–559 and 414, 560–561), with 'its dreary water' and 'ninefold circles' (1999, VI, 438–439, 562–563). Virgil transforms the Styx from river to marsh.[8]

Dante followed Virgil in transforming the Styx from river to marsh. In the seventh Canto and the fifth Circle of Dante's *Inferno* (and not the tenth level, as the Mayor of New York surmises incorrectly in *Ghostbusters II*) the condemned souls of the sullen, by a sort of sibilant (consonantal), alliterative (poetic) justice (at least in John Ciardi's English translation!), are stuck in the slime of the Styx, transformed like Virgil from river to marsh, complete with black water:

By that foul water, black from its very source,
    we found a nightmare path among the rocks
    and followed the dark stream along its course.
Beyond its rocky race and wild descent
    the river floods and forms a marsh called Styx,
    a dreary swampland, vaporous and malignant.
And I, intent on all our passage touched,
    made out a swarm of spirits in that bog
    savage with anger, naked, slime-besmutched.
They thumped at one another in that slime
    with hands and feet, and they butted, and they bit
    as if each would tear the other limb from limb.
And my kind Sage: 'My son, behold the souls
    of those who lived in wrath. And do you see
    the broken surfaces of those water-holes
On every hand, boiling as if in pain?

[8] For further discussion, see Borca (1997, 7–8).

> There are souls beneath that water. Fixed in slime
> they speak their piece, end it and start again:
> 'Sullen were we in the air made sweet by the Sun;
> in the glory of his shining our hearts poured
> a bitter smoke. Sullen were we begun;
> Sullen we lie forever in this ditch.'
> This litany they gargle in their throats
> as if they sang, but lacked words and pitch.'
> Then circling on along that filthy wallow,
> we picked our way between the bank and fen,
> keeping our eyes on those foul souls that swallow
> The slime of Hell. (Dante Alighieri 1982, 55–56)

The slime of Hell is the product of the marsh or swamp of Hell.

5.3 Monsters and Dragons

Dante's 'influence' on Milton has often been remarked, but whether this includes their similar views of wetlands and their association of hell with them is another matter for Milton saw the Styx in *Paradise Lost* as 'the burning lake,' as 'abhorrèd' and as 'the flood of deadly hate' (II, 576,7). In the same book he elaborates in more detail on Hell as a place of:

> [...] lakes, fens, bogs, dens and shades of death,
> A universe of death, which God by curse
> Created evil, for evil only good,
> Where all life dies, death lives, and Nature breeds,
> Perverse, all monstrous, all prodigious things,
> Abominable, inutterable, and worse
> Than fables yet have feigned, or fear conceived,
> Gorgons and Hydras, and Chimeras dire (II, 621–628).

This view of Hell as swamp (and of the slime as unutterable) persists through the nineteenth century and well into the twentieth century. In a similar vein 'stygian' gets overworked as an adjective in relation to the darkness and gloom of swamps. Milton used it in *Paradise Lost* to refer to 'the Stygian pool [...] that obscure sojourn' (III, 14–15).

Hell conceives and gives birth to monsters, not only Gorgons, Hydras, and Chimeras, but also Satan. Milton's ultimate monster of hell is Satan. Milton's Satan is a swamp serpent and marsh monster, or more precisely a miasmatic marsh monster. He is described as 'involved in rising mist' (IX, 75) and as 'wrapped in mist/Of midnight vapour' who glides 'obscure' (IX, 158–9) 'like a black mist low creeping' (IX, 180). Satan is a miasmatic marsh monster because as a fallen angel he has been 'mixed with bestial slime,/This essence to incarnate and imbrute' (IX, 165–6). Satan 'personifies,' or more precisely 'monstrifies,' slime and wears as his proper wardrobe the miasma which arise from it. He monstrifies slime because he was 'engendered in the Pythian vale on slime' (X, 530). He is born from slime and made of slime; he is slime by birth and slime by nature.

As a slimy swamp serpent, his mode of propulsion is not restricted to any one kind, but varies according to the (wet)landscape over which he is travelling. As he does not stick to one mode of propulsion he is aberrant: 'so eagerly the Fiend/ O'er bog or steep, through strait, rough, dense, or rare,/With head, hands, wings, or feet pursues his way,/And swims or sinks, or wades, or creeps, or flies' (II, 947–50). Satan is even a flying miasmatic marsh monster and swamp serpent, in short, a dragon. He is monstrous because he combines the physical characteristics of birds (he flies (like the Stymphalian birds)), fish (he swims), reptiles (he creeps), and quadrupeds or bipeds (he sinks or wades).

Monstrous beings embody the mixing of the four elements of earth, air, fire and water. Milton describes in more detail the watery world and womb of the first day of creation that morphs into Hell and gives birth to Satan:

> [...] this wild abyss,
> The womb of Nature and perhaps her grave,
> Of neither sea, nor shore, nor air, nor fire,
> But all these in their pregnant causes mixed
> Confus'dly, and which thus must ever fight,
> Unless th' Almighty Maker them ordain
> His dark materials to create more worlds,
> Into this wild abyss the wary Fiend
> Stood on the brink of Hell. (II, 910–918)

Milton refers to the four elements as 'the cumbrous elements, earth, flood, air, fire' (III, 715):

> [...] ye elements, the eldest birth
> Of Nature's womb, that in quaternion run
> Perpetual circle, multiform, and mix
> And nourish all things, let your ceaseless change
> Vary to our great Maker still new praise. (V, 180–184)

The mixing of elements in 'Nature's womb' does indeed nourish all things as Milton indicates. It is the wet land and warm water which give birth to life and which are the womb of the world.

God, for Milton, brings order out of the wetland chaos, out of 'the vast immeasurable abyss/Outrageous as a sea, dark, wasteful, wild' (VII, 211,2):

> [...] on the wat'ry calm
> His brooding wings the Spirit of God outspread,
> And vital virtue infused, and vital warmth
> Throughout the fluid mass, but downward purged
> The black tartareous cold infernal dregs
> Adverse to life [...] (VII, 234–239)

Hell as swamp was not just any part of hell for Milton, but the lowest part of the classical Greek underworld called Tartarus. Milton regards water as lifeless until the spirit of God breathes life into it. Milton's theology entails sublimation of the messy materiality and immanent spirituality of earth and water into the transcendental spirituality of God that has no biblical basis. Milton's anti-paludal theology needs rewriting for pro-paludal and contemporary, conservation counter-theology (as I try to do in the conclusion to the present chapter).

Swamps give birth to monsters, not only to slimy Satan, but also to swamp dragons. In the eleventh and twelfth centuries CE (so well before Milton's time) the legend of St George and the Dragon started to circulate widely in Europe (Arnold 2018, 63–70). In one version of the St George legend 'a terrifying Dragon' is an inhabitant of a lake and an instrument of God's wrath. God sent the dragon 'which devoured the inhabitants' of the nearby city ruled by 'a godless emperor' (Collins 2012, 90). St George kills the 'huge and awful' dragon with his sword.

St George is depicted killing the dragon with his sword in probably the most famous depiction of them all—Benedetto Pistrucci's design of St George and the Dragon stamped on many mintings of British gold sovereigns and half-sovereigns dating from 1817, a 1911 instance of which I wear on a gold chain as a present from my grandparents who had received it as a wedding gift. Collins (2012, 203) describes this design in his exhaustive catalogue of all things related to St George and the Dragon as 'one of the most beautiful designs ever to appear on a coin.' It may be a beautiful design, but it depicts the gruesome and horrible slaying of the dragon in a gross act of animal cruelty, as do the statues around the world discussed below. The beauty of the design hardly mitigates or overcomes the ugliness of the dragon-slaying.

For Ingersoll (2014, 194) writing in the 1920s the two main protagonists of the legend are 'the proper, most eminent Saint George, and his most celebrated and distinguished of all Dragons.' Unlike the dragon of the Old English long poem *Beowulf* who is an earth monster, the dragon of the legend of St George is a water monster, or more precisely a marsh monster, like Grendel and his mother in *Beowulf*, and/or a swamp serpent. The dragon in general for Ingersoll (2014, 15) has 'only one definable characteristic—association with and control of water.' This is certainly not the case with the dragon of *Beowulf* who has a number of defining characteristics, such as living in the earth and association with, and control of, fire.[9] In Ingersoll's summary of the St George legend 'a loathsome monster' inhabits 'a pond' of 'dark water.' Like the dragon of *Beowulf*, 'the exhalations of its breath poisoned all who came near.' St George meets 'the horrible fiend' and 'transfixed the beast' with his lance (Ingersoll 2014, 194–195). Most depictions of St George and the Dragon show him killing the dragon with his lance (Arnold 2018, 63–68). Some depictions even show St George killing the dragon with both his sword and lance, as depicted in a statue in the old city of Stockholm in Sweden, perhaps a case of overkill (Fig. 5.1).

In the thirteenth century *The Golden Legend* for Collins (2012, 91) 'popularized the legend of George and the Dragon in the West [...] The story may have been particularly well-received in England because of the legend of Beowulf in Anglo-Saxon literature' dating from the eighth century containing the English Ur-story of a dragon. The success in England of the legend of George and the Dragon was partly attributable to *Beowulf* and its marsh monsters (Grendel and his mother) and a dragon. Caxton published an English translation of *The Golden Legend* in 1483. It describes 'a stagne or a ponde lyke a see wherein was a Dragon whyche envenymed all the contre' (cited by Collins 2012, 93). 'Envenomed' is largely lost from English today (Robert Coover uses it in 'The Dead Queen'), but is certainly precise when considering the activities of the dragon of *Beowulf*. The dragon of the St George legend follows suit. The dragon in *The Golden Legend* is not merely a water monster, but specifically a swamp

[9] For further discussion of *Beowulf*, see Giblett (2018b, Chap. 3; 2021, Chap. 1).

Fig. 5.1 Statue of St George and the Dragon, Gamla Stan, the old city of Stockholm, Sweden. Photo by Rod Giblett

serpent or marsh monster. This is certainly the case for Baden-Powell, the founder of the Boy Scouts, for whom the dragon 'lived in a swamp' (cited by Collins 2012, 128). Similarly Sabine Baring-Gould's *Lives of the Saints* of 1872 recounts the legend in which 'a pond [is] infested by a monster' and how St George sees 'the monster rising from the marsh' (cited by Collins 2012, 159).

The dragon in Christianity, Collins (2012, 92) concludes, 'symbolized the Devil, heresy, paganism and ultimately anything destructive of [Christian?] society [...] Many a Dragon has the curled and twisted tail of a serpent, suggestive of evil as old as Eden.' The dragon in Jacob Boehm's statue of St George and the Dragon outside the State Library of Victoria in Melbourne has a forked tongue, which is certainly suggestive of a serpent and of deception. Whether this is also suggestive of evil as old or not as Eden is another question. Geoffrey Serle (1971, 287) might concur with Collins when he argued that this statue signifies 'the triumph of good over evil' (Fig. 5.2).

Fig. 5.2 Jacob Boehm, Statue of St George and the Dragon, State Library of Victoria, Swanston St, Melbourne, Victoria, Australia. Photo by Rod Giblett

Yet the dragon as a creature of primeval chaos as depicted in *Genesis* 1: 2 may be older than Eden, and may not be suggestive of evil at all. A twisted tail and forked tongue are suggestive of a serpent, but are not necessarily suggestive of evil, as old as Eden or not, and only with the hindsight of the biblical story of the fall.

Collins (2012, 93) goes even further to suggest that the dragon's 'poisonous breath may be taken to be heretical preaching.' Unlike the dragon's poisonous breath or venomous vapour that could kill (as in *Beowulf*), heretical preaching cannot, though it might get the heretical preacher burnt at the stake so in a sense it could kill—by fire, like the dragon's fiery breath. The dragon was a salutary warning that heretical preaching was dangerous and could kill.[10]

5.4 Site of Resistance

The marsh is not only a refuge for the vulnerable and a womb for new life, dragons and other monsters, but also a site of resistance, especially for indigenous peoples against imperial invaders, such as the Romans in England. English wetlands have, as Jeremy Purseglove (1989, 35) has pointed out, 'a long history as centres of resistance.' This long history stretches from the ancient Romans and Britons, through the Danes and Alfred the Great to the Normans and Hereward the Wake. In many ways this has been a predictable history because, as Purseglove goes on to observe, 'marshes have always been easy to defend.' And hard to attack. That defence and long history first makes its way into the written record with the Roman invasion in which H. C. Darby (1974, 20) noted that the Romans 'encountered much difficulty' in their conquest of marshy regions. Xiphilinus related how in peace the Britons 'inhabit mountains wild and waterless, and plains desert and marshy' and how in war they are capable of 'hiding in marshes [...] many days with heads only out of water' (cited by Darby 1974, 20). The Romans, on the other hand, according to Dion Cassius 'wandered into the pathless marshes, and lost many of their soldiers' (cited by Darby 1974, 20).

[10] For animal-friendly and pro-paludal retellings of the second part of *Beowulf* involving a dragon and of the St George and the Dragon legend, see Giblett (2018a, c, 2019a, c).

An embattled indigenous population on its home 'turf' of a wetland was an intractable opponent as the Americans and their allies were to rediscover in the Plain of Reeds and the Mekong Delta in Vietnam. This was also particularly so with the Fens in England whose 'inhabited fields were isolated, their waters tidal, with great banks of treacherous mud, intricate and unbridged; such conditions are amply sufficient for a defensive war' as Hillaire Belloc (1970, 53) surmised. Belloc (1970, 158) maintains that 'the main cause of the breaking down of the Roman Road was marsh.' Could wetlands (the Pontine Marshes, the marshes of the Teutoburger Wald, the Fens and other English marshes) even be a cause for the breaking down of the Roman Empire? In the first century Tacitus (1970, 136–137) seemed to have the Teutoburger Wald in mind when he described Germania as 'a land of bristling forests and unhealthy marshes (*paludibus foeda*).'[11] Could wetlands be the main cause of the breaking down of any empire for that matter? Paddy fields, the Mekong Delta and the Plain of Reeds certainly helped to defeat American imperialism in Vietnam.[12] Wetlands defy empire.

In cataloguing 'the more important physical factors modifying the formula of the Road' a century ago Belloc mentions marshes first and foremost and then goes on to suggest that:

> it is not always appreciated that the chief obstacle to travel from the beginning of time has been and still remains marsh, which may be defined as soil too sodden for travel, as distinguished from the lands which are boggy in wet weather but passable. Marsh is less striking to the eye, especially to the modern eye, than a stretch of water, much less striking than the apparent obstacle of the sea, or of a bold hill range: it is nevertheless the chief problem presented to the making of a road, because of all natural obstacles *it is the only one wholly untraversable by unaided man* [Belloc's emphases]. Man [*sic*] unaided can climb hills, swim water, work his way through dense undergrowth. But marsh is *impassable* to him: it is *the great original obstacle to progress* [my emphases]. (Belloc 1923, 13–14)

Belloc shifts quickly from marsh as obstacle to travel to marsh as offensive to the eye, from engineering to aesthetics, both of which are

[11] The editor of the Loeb Classical Library edition of *Germania* mentions the Teutoburger Wald in a footnote at this point (Tacitus 1970, 136n2). For further discussion of the Teutoburger Wald, see Schama (1995, 88–91).

[12] For further discussion of wetlands as sites of resistance, see Giblett (1996, Chap. 9).

nevertheless intimately associated in modernity, albeit in a merely functional hierarchical way, as we saw in the previous chapter in *The African Queen* in which 'no place could be beautiful that presented navigational difficulties' (Forester 1956, 135).

The response of 'progress' to the obstacle of the marsh has been to avoid it wherever possible ('our modern ways avoid marsh' (Belloc 1923, 15)), or to fill it or drain it where avoidance has not been possible, or more precisely where modern technology has made avoidance unnecessary, as in the case of cities such as St Petersburg or Perth (Western Australia). Indeed, if the wetland is the chief obstacle to progress as Belloc contends, then drainage is the epitome of 'Progress' (Davison 1993, 103 and 115). In the process the land is reduced to surface to be travelled over or built upon with any impediment to either being constituted as an obstacle, not only to travel or cities, but to 'progress,' that motto of capitalism. The modern city, however, is characterized by *not* being compelled to avoid marshes. Modern technology, the engineering of pumps, shovels and canals, made it possible for cities to go where no city had ever gone before, that is, into wetlands albeit after they had been drained or filled. Modern ways and the marsh are totally and irretrievably inimical to each other. They cannot co-exist; the marsh must go to make way for the modern.

Conversely, the return of the marsh for Belloc constitutes a return to the barbarian: 'whenever civilization breaks down you begin to get a series of marshes, with all their accompaniments of fever and the rest [whatever that is, something horrific and unutterable anyway] growing up along the roads' (Belloc 1923, 65). For Belloc, roads and drains are the bastions of modern industrial and urban civilization holding out against the incursions of 'primeval marsh' (Belloc 1923, 104).

The marshy conditions in the Fens were more than amply sufficient in the defensive war against the Romans who, according to Mary Chamberlain (1975, 15):

> considered the area ungovernable and were outwitted by the cunning fen dwellers who could travel on stilts across the marshes, a skill which the Romans failed to master. The inaccessibility of the area made it a refuge for those fleeing from authority. The poverty and atmosphere of independence made it ripe for revolution and a stronghold of anti-authoritarianism.

The Fens were the site of a long tradition of British resistance whose, perhaps the most exemplary, figure was Hereward the Wake.

In his 'Prose Idyll' of the Fens Charles Kingsley (1884, 111) describes:

> how Nature left to herself, ran into wild riot and chaos more and more; till the whole fen became one 'Dismal Swamp,' in which the 'Last of the English' (like Dred in Mrs Stowe's tale) took refuge from their tyrants, and lived, like him, a free and joyous life awhile.

The 'Last of the English' is Kingsley's appellation for Hereward the Wake to whom he devoted a whole book. *Hereward the Wake* exemplifies Kingsley's ambivalence about the Fens encountered in the previous chapter of the present volume. The Fens for Kingsley (1908, 9), 'have a beauty of their own, these great fens, even now, when they are dyked and drained, tilled and fenced—a beauty as of the sea, of boundless expanse and freedom.' It is this feature of the Fens in particular, the way in which for Kingsley (1908, 9) 'the green flat stretched away, illimitable,' which gives the Fens a large horizon of possibilities, especially for defence. In the Fens in particular and in the marsh in general, using Belloc's words, 'one of the first requirements of defence is afforded—an unbroken view of every avenue by which attack can come. There is no surprising such forts' (Belloc 1970, 53–54).

Such a fort was the Isle of Ely which Kingsley describes as 'a natural fortress' (Kingsley 1908, 9). The precise feature which makes the Fens such a good defence for the Fen defender and such an object of aesthetic appreciation for the Fen dweller is precisely, and hardly surprisingly, the feature which the stranger bemoans, even vilifies. Yet for Kingsley despite its obvious aesthetic and military virtues, 'that fair land, like all things on earth, had its darker aspect. The foul exhalations of autumn called up fever and ague, crippling and enervating' (1908, 10). Against the beauty, even sea-like sublimity, of the fens Kingsley poses the miasmatic and melancholic diseases arising from its slime.

The lowlands generally, including the wetlands presumably, for Kingsley (1908, 1 and 2) are curiously on the one hand 'the soonest to be civilised and therefore the soonest taken out of the sphere of romance and wild adventure into that of order and law, hard work and common sense,' and so into the imperialist order of the straight lines of ditch, drain, dyke and road, and on the other, the place of the backward and primitive: 'the lowlands and those who live in them are wanting in the poetic and romantic elements,' another case of 'blaming the victim,' the victim of 'civilization,' of drainage and enclosure.

The lowlands also contrast unfavourably with the highlands, as the lowlands for Kingsley (1908, 2) lack the mystery and sublimity of the highland: 'there is in the lowland none of that background of the unknown, fantastic, magical, terrible, perpetually feeding curiosity and wonder, which still remains in the Scottish highlands.' Yet Kingsley (1908, 10) describes only a few pages later how the Fens were the home of 'witches, ghosts, Pucks, Wills o' the Wisp.' This indigenous, pre-modern tradition of the fantastic and magical which fed curiosity and wonder in a satisfying diet for millennia in an autochthonous culture is counterposed unfavourably to the modern, Romantic sense of the sublime.[13] Lowlands and highlands are counterposed, as are implicitly wetlands and drylands, and the high- and dry-lands are valorised over and to the detriment of the wet- and low-lands, the nether-lands (see Giblett 1996, Fig. 1, Chap. 2).

The dry highlands are the locus of the sublime whereas the wet lowlands are the site of slime, a feature which Kingsley dwells upon repeatedly in *Hereward the Wake*, especially when the slime is ostensibly 'bottomless' (Kingsley 1908, 282 and 286) and certainly 'stagnant' (421). The bottomlessness of slime contrasts unfavourably with the heights of the sublime. Yet it is this very same slime that is no less than 'the camp of refuge for English freedom' (11) and the ultimate defence against the invading Normans who end up, according to Kingsley, 'drowning in the dark water,' the black waters of malaria and melancholy, 'or more hideous still, in the bottomless slime of peat and mud' (286), a fate worse than, and on top of, or more precisely beneath, death.

In a psychogeocorpography of modernity the seemingly bottomless and unplumbable depths of the slime of the wet lowlands (the interiority of the Body of the Mother) contrasts unfavourably with the immense but measurable, terrifying but scalable, heights of the sublime of the dry highlands (the exteriority of the Law of the Father) (see Giblett 1996, Fig. 1, Chap. 2). Similarly the clear, running, white water of the mountain stream conducive to bracing clarity of thought and flowing in demarcated watercourses contrasts favourably with the muddy, stagnant, black water of the wetland lying beneath the black sun of depression and melancholia and sitting still in labyrinthine backwaters. Parts of the Fens are even called 'Black fen,' 'black for the dark peaty soil; black—for the mood of the area, for its history and for its future,' as Mary Chamberlain explains (1975, 11), even black for its waters.

[13] For further discussion of the sublime, see Giblett (1996, Chap. 2).

Rather than a staging ground for rebellion coming out of the wetland against a centralized authority, the wetland has been largely a site of resistance against the incursions and depredations of that authority into the wetland. Darby (1974, 143) notes that:

> there were many local rebellions of both political and of economic origin in medieval England; but it is significant that the people of the fens never took part in any of these [...] Their country certainly became a centre of resistance to authority, but to interpret that fact as the expression of any spirit of freedom fundamental in marshland peoples is quite unwarranted [...] The role of the Fenland in English history has been passive rather than active. It was repeatedly a region of refuge, rather than a breeding ground of freedom and discontent [...] the problem of fenland rebellion is not a social one with an economic background, but a military one with a topographical background.

Local rebellions of both political and economic origin did take place in the Fens in early modern England. The problem of early modern fenland rebellion is a social one with an economic background with a unique topography, especially with the enclosure of the Fen commons into private property (as we will see in the following chapter of the present volume).

In Paul Kingsnorth's *The Wake* published in 2014 the Fens are a place of burgeoning and decomposing life, of both life-giving and death-dealing water, of sacrality and resistance to hegemony. Written in what Kingsnorth calls, and devised as, 'shadow English,' 'a pseudo-language' that might have been that adumbrates Old and Modern English, *The Wake* is an historical novel set in the aftermath of the Norman invasion of England in 1066, and geographically in and around the Fens. For Buccmaster, the first-person narrator of *The Wake,* 'the fenn' 'was what has macd us who we is as folc this place' (Kingsnorth 2014, 33). The shallow waters of the wetland are fascinating and delightful as they are referred to as 'the deorc undeop waters of the fenn' (34). The dark waters of the wetland are not an object, or abject to use Kristeva's (1982) more apt term, of horror that hides lurking monsters. Buccmaster acknowledges that 'the fenn that can be so blaec and deop and cold on this mergen was a thing of great wundor' (50). Buccmaster has heard the Beowulf story. When he is a child he goes out in his grandfather's boat into the fen and fears that the world under the water is where Grendel and his mother were living and that they 'was cuman for me' (51). The world under the water is the home, not to monsters, but to the old sacrality and folk knowledges. It is also the site of resistance to the Normans as it was for the historical Hereward of the same period usually known as 'the wake,' but Kingsnorth seems to be implying or suggesting that Buccmaster is another wake, or the real wake.

The fens are not necessarily a pleasant place, but they are subject to commonplace misconceptions. Buccmaster relates 'it is saed that the fenns is all deorc waters and deope mud' and 'that folcs of the fenns' is 'yfel wights of the mere.' The fens are not the place of evil monsters. Buccmaster counters these misconceptions:

> These is the things folcs saes who does not cnawan naht of what they is specan for the fenns is a place of wundor to those who cnawan them. yes there is deop meres and waters so blaec that oxen is lost in them and nefer seen and there is muds what strecces for miles by secg and lesch [reeds]... there is the blaec fenns where the eorth and the waters in all blaec like the graef. (Kingsnorth 2014, 123)

Buccmaster acknowledges the unpleasant aspects of the Fens, including their black water.

Buccmaster initially seeks shelter from the Normans in the wood but later comes to the realization that 'i was a man of the water not the wud the fenns was callan me' and that he must go to 'the water lands for there was no other place triewe in this world' (Kingsnorth 2014, 286). When he returns to the Fens it is a return to his home land with familiar black earth, meres and sedges and 'in the heofon was the eald feel of my life and of my folc' (287). He acknowledges that 'i missed the fenns' and that 'it is the triewest place in angland' (287). The Fens are the quintessential Eng-land, the English home land, the repository of sacrality (as we will see in the following chapter of the present volume).

5.5 The Swamp as Hell, or Heaven

What Bruce McGarvey (1988, 59 and 61) calls 'a myth of the hellish swamp' and 'myths of the swamp as Hell' have been around in western culture for a long time—myths of hell as the swamp circulated in ancient Greek and Roman cultures, Medieval Italian culture and English Renaissance and Reformation culture (as we have seen in the present chapter). More recently, in modern western culture, myths of the swamp as hell surfaced. The view of the wetland as hell is writ large in Peter Matthiessen's

Killing Mister Watson published in 1990. Set in the Florida Everglades, this factive novel can be read as an attempt to deconstruct the myth of the swamp as hell from within through its play of narratorial voices; yet in the process it reproduces some other myths such as the psychogeopathological association between the anal, the excremental and the aquaterrestrial. One of the narrators, Henry Thompson, wonders of one family of early settlers:

> with all of Florida to choose from, who else would come to these overflowed rain-rotted islands with not enough high ground to build an outhouse, and so many skeeters plaguing you in the bad summers you thought you'd took the wrong turn to Hell. (Matthiessen 1990, 10, 11).

When hurricanes wreak havoc across this wetlandscape they do not merely reveal the brown 'under-raiments' beneath the normally placid surface of the fatal margins (hardly a shore). Rather, they overturn the privies and turn up all that is rotting beneath the surface:

> in the hurricane's wake, the labyrinthine coast where the Everglades' deltas meet the Gulf of Mexico lies broken, stunned, flattened to mud by the wild tread of God [...] In the dark air a sharp fish stink is infused with corruption of dead animals and blackened vegetables, of excrement in overflowing pits from which shack privies have been blown away (Matthiessen 1990, 1, 2).

This is hell because God has trodden on it and damned it to destruction; these are, according to another narrator in the novel, 'godforsaken swamps' (Matthiessen 1990, 214) though previously the same narrator had claimed that some 'used to call this place God's country, and we still do, cause nobody but God would want no part of it' (Matthiessen 1990, 56)

The swamp is hell not only because God has forsaken it and trodden upon it, but also because it is the place of timelessness, the place where time seemingly stands still. In Vereen Bell's 1941 novel *Swamp Water* set in the Okefenokee Swamp of Georgia:

> the swamp was a lost world; a nether land where the crawling things in the muck and the screaming things in the air had triumphed, and man rotted in the peaty earth; the space was the space of eternity, endless, changeless. (Bell 1981, 212)

The wetlandscape is represented here as the landscape of eternity, the space of timelessness; wetlandscape as endless timescape (see Giblett 1996, Fig. 1, Chap. 2). Ecologically this is nonsense as the wetland is arguably the most dynamic and most mutable of all ecosystems. Rather than being timeless, wetlands are 'timeful.' Wetlands are the space of mutability. David Rains Wallace (1987, 207) has pointed out that 'wetlands are intermediaries between past and future as well as between water and land.' In fact, the wetland serves a vitally important function to constantly change. If it does not change, a wetland dies. A static wetland is a dead wetland.

The wetland is the space of eternity not because it is changeless but because it is ever-changing, ever life-reproducing like the placenta and life-restoring like blood. The only constant is change, and the wetland is the exemplar of this principle. As Wallace (1987, 216) goes on to summarize:

> various studies have indicated that not only that salt marshes produce most of our seafood, but that swamps store water with much less evaporation than reservoirs, that marshes and swamps purify sewage effluent more cheaply than treatment plants, and that floodplain forests regulate stream flow more cheaply than flood-control reservoirs. Wetlands are not simply living museums for the preservation of alligators and storks, but a vital part of the hydrologic system that produces clean and reliable water supplies.

Yet those wetlands which are habitats for alligators and storks, crocodiles and jabirus, alone and do not have any immediately exploitable part to play in the hydrologic system should still be preserved. Wetlands should not be preserved just for their limited, utilitarian benefits to humans, but also for their benefits for other species and for their aesthetic, ecotourist, 'passive' or eco-recreational and spiritual benefits to humans.

The wetland is integral to the evolutionary process, and so it defies or upsets the conventional reading of the Judeo-Christian creation story. Referring to the Black Swamp of Ohio, one early colonial settler remarked that 'we read that God divided the land from the water, but here is a place He forgot' (cited by McGarvey 1988, 67). God has been seen to have 'forgotten' or forsaken wetlands. When Dr Watson entered Great Grimpen Mire in Conan Doyle's *The Hound of the Baskervilles* published in 1902 he reported to Sherlock Holmes from 'this most God-forsaken corner of the world' that he had 'left all traces of modern England behind' (cited by Giblett 2019b, 103–104). He had time-travelled to the pre-modern past when the world was young and God was in the process of creation. When

God divided the land from the water, He did not divide all land from all water. He left wetlands undivided. Wetlands as areas between, or both, land and water, slipped through. God created everything and so created wetlands. But as He did not completely divide all land from the water, wetlands were not only an aberration but also a vestige of, or a throwback to, the first day of creation when everything was chaotic, disordered, and the spirit of God brooded on the face of the waters.

The Ohioan colonial settler was a good reader of the Bible as s/he paraphrases *Genesis* 1: 6. It states that 'God said, Let there be a firmament in the midst of the waters, and let it divide the waters from the waters.' In the seventeenth century William Dugdale (1662) wilfully misquoted this verse to justify after the fact the large-scale drainage of the English Fens were a God-given task when he wrote as the opening gambit of his history 'that works of Drayning are most antient, and of divine institution, we have the testimony of holy Scripture. 'In the beginning God said, let the waters be gathered together, and let the dry land appear; and it was so'.' In the beginning God created wetlands (*Genesis* 1: 1). In order to separate land and water, they were not separated in the beginning. In separating land and water, God did not drain water from the land.

Ash (2017) gives a recent detailed historical account of the draining of the Fens in the seventeenth century, including a discussion of Dugdale's role in documenting and legitimizing it. Ash states in passing that 'draining the Fens [...] was God's manifest will' and later that 'the first drainage projector in his [Dugdale's] account was God, who had had separated the earth from the water at the Creation' (Ash 2017, 182–183, 291). At the creation, on the first day, God did not separate earth and water; that was on the second day, after the first creation of the heavens and the earth. Arguably separation is not creation, but destruction.

In an earlier article Ash (2016, 456) argued that 'a few projectors and their supporters viewed fen drainage as [having] a divine mandate.' Dugdale was among the supporters who misquoted the bible and used it to legitimize the projectors' own capitalist, political and economic power and profit-making, a not uncommon practice then and since. Yet this was all justification after the fact as earlier royals, royalists and republicans alike generally did not reach for their bibles to legitimize their own capitalist, political and economic power and profit-making in draining the Fens. The enclosure of the commons into private property was 'a plain enough case of class robbery,' as the late, great E. P. Thompson (1968, 237) said (discussed in the following chapter of the present volume). Ash neglects

Thompson's critique and the class politics and capitalist economics of the draining of the fens and enclosing of commons into private property, opting instead for euphemistic 'propertied elites,' etc., including royalty.

King Charles I played a decisive and deleterious role in the draining of the Fens and enclosing commons, a fact not lost on some Fenlanders. Ash relates how the Epworth rioters in 1628 'identified King Charles as the main source and target of their discontent' as he 'had personally concluded the agreement with [Cornelius] Vermuyden to drain *their lands*' (my emphasis; Ash 2017, 142). Ash later argues that 'draining the Fens, Charles [I] believed, was an intrinsic good that would benefit the commonwealth and boost royal revenues through improvement' (Ash 2017, 175). The result was that 'virtually all the major English projects undertaken in the seventeenth century were begun after Charles came to the throne in 1625' (Ash 2017, 175). In November 1641 the Commons presented to Charles I a set of grievances known as 'the Grand Remonstrance', including the complaint that 'large quantities of common [...] grounds hath been taken [...] by colour of the Statue of Improvement' (Ash 2017, 205). The long battle over its passage in parliament between divided royalists and parliamentarians led to the Civil War and the execution of Charles as a tyrant in 1649 (Ash 2017, 205–206, 215). Aquaterracide led to regicide.

Yet killing off the king did not lead to the cessation of draining the Fens, quite the contrary. The Civil War, Ash goes on to argue:

> had wrought many changes on England's political landscape, but it had not altered the plight or fate of the Fens. [...] The new Commonwealth regime was just as enthusiastic about draining the Fens and profiting from the enclosure of [common wetlands into] improved [dry]land [and private property] as the Crown had been. (Ash 2017, 217, 219).

The aquaterracide led by Charles I from 1625 to 1649 was perpetuated by the Commonwealth of 1649–1660. Royalty and republicans alike were drainers. Especially galling for many Fenlanders was the fact that they had fought on the parliamentarians' side against the royalists in the Civil War in the mistaken belief that the parliament would support and prosecute their resistance to the draining of the Fens. The aquaterracide led by Charles I and continued by the Commonwealth was perpetuated in the British empire, including Australia. Could Charles III right some of the historical wrongs of Charles I and make some amends for them by instituting a programme rehabilitating wetlands in the UK and Commonwealth countries along the lines of 'the Queen's Green Canopy' in the UK and 'the Queen's Commonwealth Canopy' that plants trees? 'The King's Commonwealth Wetlands'?

The wetland harks back to the time when the entire earth was wetland. *Genesis* 1:1 needs to be rewritten or retranslated in postmodern ecological terms which recognizes the importance, even primacy, of wetlands: 'in the beginning God created the heavens and the earth, and the earth and the heavens were wetlands.' But perhaps when God divided partially the land from the water it was not so much that He forgot or forsook wetlands than that He knew that the land and the water and the creatures that He later created to live in or on them would not survive and would not be able to reproduce without wetlands. Subsequent days of creation do not supersede and 'forget' about their predecessors but build and rely upon them. In the Judeo-Christian creation story the days have been sees in terms of a progressive, supersessive time-line heading towards the ultimate goal of humanity, and not as an evolutionary process.

Without the first day of creation, the creation of the earth as wetland, there is no sixth day, no creation of humankind. Without the creation of wetlands on the first day, and every day for that matter, there is no recreation of human life, there is no human life full stop, on the sixth or any other day. Rather than being forgotten or forsaken by God, the wetland is the first day of creation, God's first, and for many best, work, without which no other day could ever begin, could ever dawn. Rather than being hell, the wetland is heaven because it was created at the same as the earth. When one goes to the wetland one goes back to the first day of creation when it springs fresh and new from the hand and word of God, when the spirit broods over the earth as wetland like a hovering marsh harrier, not like a malevolent ghost or swamp monster

The swamp, or the wetland more generally, is a place of death. But it is also a place of life. Wetlands are places of both life and death. They are living black waters where, as David Miller (1989, 80) puts it, 'the intermingling of life and death' takes place. Delia Owens has expressed this intermingling elegantly and eloquently in the prologue to her recent novel, *Where the Crawdads Sing*:

> the marsh is not swamp. Marsh is a space of light, where grass grows in water, and water flows into the sky. Slow-moving creeks wander, carrying the orb of the sun with them to the sea [...] Then within the marsh, here

> and there, true swamp crawls into low-lying bogs, hidden in clammy forests. Swamp water is still and dark, having swallowed the light in its muddy throat [...] There are sounds, of course, but compared to the marsh, the swamp is quiet because decomposition is doing its cellular work. Life decays and reeks and returns to the rotted duff; a poignant wallow of death begetting life [...] A swamp knows all about death, and doesn't necessarily define it as tragedy, certainly not a sin. (Owens 2018, 3–4)

The swamp is heaven and hell, or neither heaven nor hell. The swamp is its own place, complete and sufficient to itself where life and death intermingle.

Along similar lines and in this regard, the anti-paludal sections of Milton's *Paradise Lost* cited and discussed previously in the present chapter need to be rewritten for a pro-paludal and contemporary conservation counter-theology in a poem that could be called 'Her Dank Materials':

> Earth, air, fire and water:
> The unencumbered elements,
> The eldest birth of Nature's womb,
> That in four ways run perpetual circle,
> Multiform and mix and nourish all things,
> Let your ceaseless change
> Sing to our great Mater still new praise.
>
> Lakes, fens, bogs, dens and spirits of life,
> A universe of life, which God by love
> Created good, only for good, not evil
> Where all life dies, new life lives, and Nature breeds,
> Promiscuous, all maternal, all prodigious things,
> Fascinating, horrifying, and better
> Than fables yet have feigned, or fear conceived,
> Marsh monsters, swamp serpents and dragons fiery.
>
> This wild swamp,
> The womb of Nature and perchance her tomb,
> Of foursome sea, and shore, and air, and fire,
> With all these in their pregnant elements mixed
> Creatively, and which thus must ever in concord,
> That the All-making Mater them ordain

Her dank materials to create more worlds,
Into this wild swamp the wary friend
Stood on the brink of Heaven.

On the wat'ry calm
Her brooding wings the breath of wind outspread,
And vital virtue suffused, and vital warmth
Throughout the fluid mess, so outward splurged
Her black, warm, slimy dank materials
Conduce for life.

Amen—so be it.

References

Arnold, Martin. 2018. *The Dragon: Fear and Power*. London: Reaktion Books.

Ash, Eric H. 2016. Reclaiming a New World: Fen Drainage, Improvement, and Projectors in Seventeenth-Century England. *Early Science and Medicine* 21: 445–469.

———. 2017. *The Draining of the Fens: Projectors, Popular Politics, and State Building in Early Modern England*. Baltimore, MD: Johns Hopkins University Press.

Bell, Vereen. 1981. *Swamp Water*. Athens, GA: University of Georgia Press. First published 1941.

Belloc, Hillaire. 1923. *The Road*. Manchester: Charles W. Hobson.

———. 1970. *Hills and the Sea*. Westport, CT: Greenwood Press. First published 1906.

Borca, Federico. 1997. Palus Omni Modo Vitanda: A Liminal Space in Ancient Roman Culture. *Classical Bulletin* 73 (1): 3–12.

———. 2000. Towns and Marshes in the Ancient World. In *Death and Disease in the Ancient City*, ed. Valerie M. Hope and Eireann Marshall, 74–84. London: Routledge.

Chamberlain, Mary. 1975. *Fenwomen: A Portrait of Women in an English Village*. London: Virago.

Cicero. 1972. *The Nature of the Gods*. Trans. H. McGregor. London: Penguin.

Collins, Michael. 2012. *St George and the Dragons: The Making of English Identity*. London: CreateSpace.

Dante Alighieri. 1982. *The Inferno*. Trans. J. Ciardi. New York: Signet Classics. First published in Italian in the fourteenth century CE.

Darby, H.C. 1974. *The Medieval Fenland*. 2nd ed. Newton Abbot: David and Charles.
Davison, Liam. 1993. *Soundings*. St Lucia: University of Queensland Press.
Dugdale, William. 1662. *The History of Imbanking and Drayning of Divers Fenns and Marshes, Both in Forein Parts and in this Kingdom, and of the Improvements Thereby Extracted from Records, Manuscripts, and Other Authentick Testimonies*. https://quod.lib.umich.edu/e/eebo/A36795.0001.001?view=toc. Accessed 16 Sept 2022.
Forester, C.S. 1956. *The African Queen*. Harmondsworth: Penguin. First published 1935.
Giblett, Rod. 1996. *Postmodern Wetlands: Culture, History, Ecology*. Edinburgh: Edinburgh University Press.
———. 2011. *People and Places of Nature and Culture*. Bristol: Intellect Books.
———. 2016. *Cities and Wetlands: The Return of the Repressed in Nature and Culture*. London: Bloomsbury.
———. 2018a. *The Dragon and Saint George: A Fairy Tale Novella*. Cambridge: Pegasus/Elliott Mackenzie.
———. 2018b. *Environmental Humanities and Theologies: Ecoculture, Literature and the Bible*. London: Routledge.
———. 2018c. *Tales of Two Dragons*. London: Austin Macauley.
———. 2019a. *Black Swan Saga*. Shahjah: Austin Macauley.
———. 2019b. *Environmental Humanities and the Uncanny: Ecoculture, Literature and Religion*. London: Routledge.
———. 2019c. *The Oxbridge Book of Dragons*. London: Olympia.
———. 2021. *Wetlands and Western Cultures: Denigration to Conservation*. Lanham, MD: Lexington Books.
Ingersoll, Ernest. 2014. *Dragons and Dragon Lore*. New York: Cosimo Classics. First published 1928.
Kingsley, Charles. 1884. The Fens. In *The Works, Volume XV, Prose Idylls*, 93–131. London: Macmillan.
———. 1908. *Hereward the Wake*. London: Dent. First published 1866.
Kingsnorth, Paul. 2014. *The Wake*. London: Unbound.
Kristeva, Julia. 1982. *Powers of Horror: An Essay on Abjection*. Translated by Leon Roudiez. New York: Columbia University Press. First published in French in 1980.
Longinus. 1965. On the Sublime. In *Classical Literary Criticism*, 99–158. Trans. T.S. Dorsch. Harmondsworth: Penguin.
Lucretius. 1994. *On the Nature of the Universe*. Trans. R.E. Latham, revised by John Godwin. London: Penguin.
———. 2007. *The Nature of Things*. Trans. A.E. Stallings. London: Penguin.
Matthiessen, Peter. 1990. *Killing Mister Watson*. London: HarperCollins.

McGarvey, Bruce E. 1988. Landscape Myths of the Black Swamp, Part One. *Northwest Ohio Quarterly* 60 (2): 57–68.

Miller, David. 1989. *Dark Eden: The Swamp in Nineteenth-Century American Culture*. Cambridge: Cambridge University Press.

Ovid. 1984. *Metamorphoses*. Trans. F.J. Miller. Revised by G.P. Goold. Loeb Classical Library 42. Cambridge, MA: Harvard University Press.

Owens, Delia. 2018. *Where the Crawdads Sing*. Boston: Little Brown.

Pliny the Elder. 1942. *Natural History: Books 3–7*. Trans. H. Rackham. Loeb Classical Library 352. Cambridge, MA: Harvard University Press.

———. 1991. *Natural History: A Selection*. Trans. J.F. Healy. London: Penguin.

Purseglove, Jeremy. 1989. *Taming the Flood: A History and Natural History of Rivers and Wetlands*. Oxford: Oxford University Press.

Schama, Simon. 1995. *Landscape and Memory*. London: HarperCollins.

Serle, Geoffrey. 1971. *The Rush to be Rich: A History of the Colony of Victoria, 1883–1889*. Carlton: Melbourne University Press.

Sofoulis, Zoë. 1988. Through the Lumen: Frankenstein and the Optics of Re-Origination. Ph.D. Thesis, Santa Cruz: History of Consciousness, University of California.

Tacitus. 1970. *Agricola, Germania, Dialogues*. Trans. M. Hutton, W. Peterson, revised by R.M. Ogilvie, E.H. Warmington, M. Winterbottom. Loeb Classical Library 35. Cambridge, MA: Harvard University Press.

Thompson, E.P. 1968. *The Making of the English Working Class*. Harmondsworth: Penguin.

Tinarelli, Antionio. 2006. Rice in Western Culture. In *Chiccodoro: Il riso nutrizione e salute*, ed. Attilio Giacosa, Mariangela Rondanelli, and Antonio Tinarelli, 17–28. Certosa di Pavia, Italy: Torchio de' Ricci.

Varro. 1935. *On Agriculture*. Trans. W.D. Hooper, revised by Harrison Boyd Ash. Loeb Classical Library 283. Cambridge, MA: Harvard University Press.

Virgil. 1999. *Eclogues, Georgics, Aeneid, Books 1–6*. Trans. H.R. Fairclough. Revised by G.P. Goold. Loeb Classical Library 63. Cambridge, MA: Harvard University Press.

Wallace, David Rains. 1987. *Life in the Balance: Companion to the Audubon Television Specials*. San Diego, CA: Harcourt Brace Jovanovich.

CHAPTER 6

Fen Britons

Recent work on the English Fens coming out of art history, landscape aesthetics and landscape architecture demonstrates the still contentious and problematic positioning of the Fens in English culture, the need for concerted efforts to conserve them and the value of recent work to restore them (Hillsdon 2020). Vittoria Di Palma's (2014) lavishly illustrated and eponymously titled art history of wasteland devotes a chapter to the English Fens that critiques their enclosure and draining, appreciates their unpicturesque representation in landscape painting and traces the development of artificial wetlands in landscape architecture. Firmly located within the modalities of visual representation and the disciplines and discourses of art history and landscape aesthetics and architecture, this study neglects or overlooks the miasma, malaria and melancholy of the Fens as diagnosed and discussed in medicine, philosophy and psychology. These topics and disciplines are included in a more transdisciplinary approach to the Fens engaging with cultural, historical and literary studies in the environmental humanities beginning with *Postmodern Wetlands* (Giblett 1996).

Other disciplines, such as archaeology, and other arts, such as church architecture and construction (including its legendary carpentry), are extensively and thoroughly considered by Francis Pryor (2019) in his personal and professional memoir of 40 years of study of the Fens' built structures and agricultural practices in ancient, mediaeval and modern times. Recently the Fens have become the focus and site of 're-wetting' and

R. Giblett, *Wetland Cultures*, Literatures, Cultures, and the Environment, https://doi.org/10.1007/978-3-031-57365-1_6

recreating 'paludiculture' with 'a 100-year plan to re-wet 3,700 hectares of peat across Cambridgeshire,' plant more than 200,000 plugs or tube-stocks of commercially viable native grasses, rushes and mosses, halt the loss from the Fens of '4.5 million cubic metres of peat a year to soil erosion' and conserve water in the land for farming and wildlife. It is 'one of the largest restoration projects of its kind in Europe' (Hillsdon 2020).

Building on *Postmodern Wetlands*, drawing on Di Palma's work and interweaving Pryor's recent account of his work on the Fens with discussion of the depiction of the Fens in paintings by John Sell Cotman, Peter de Wint and 'Fenland artist' Nick Tearle (n.d.) and verse by Fenland poet Becky Owen-Fisher (2017), the present chapter takes a radical political approach to the Fens and the enclosure of the commons in private property, as critiqued by John Clare, Karl Marx, E. P. Thompson and Raymond Williams. It also takes a cultural, literary and historical approach by considering the depiction of the Fens in painting and the literature of the Fens, beginning with Fenland native Clare's poetry and prose in the nineteenth century for the Fens and against enclosure. It then proceeds to a consideration of the Fens in later prose, beginning with Dorothy Sayers' classic detective novel *The Nine Tailors* (1934), Graham Swift's wetland-mark *Waterland* (first published in 1983) and his 25th anniversary introduction of 2008, Phillip Pullman's *Northern Lights* (1995), the first part of his trilogy, *His Dark Materials*, Robert Macfarlane's (2008) 'new nature writing' of 'Ghost Species,' culminating recently with Becky Owen-Fisher's poetry (2017) and Stella Tillyard's historical romance *The Great Level* (2018).

6.1 The Fens in Art History

In her chapter on the Fens in *Wasteland*, Di Palma (2014, 95) argues that 'the swamp is a landscape that is dangerous not only to the body, but also to the soul.' The Fens were considered dangerous to the body because of their bad airs (as Di Palma discusses, but she does not go on to point out the reason for this view). The Fens were considered dangerous according to the miasmatic theory of disease that contended with the germ theory of disease in medical theory and practice up to the late nineteenth century. The swamp is dangerous to the body because ague (or malaria—literally 'bad air') was thought to be contracted by breathing the miasma or bad air rising from swamps, marshes and other stagnant pools of water, including in cities, especially slums (as we saw on both counts in previous chapters of the present volume). Nor does Di Palma discuss the disease of ague (or malaria) that was commented upon by many writers who bemoaned the unhealthiness of the Fens and other wetlands around the world in accordance with the miasmatic theory of disease, instances of which could be enumerated almost *ad infinitum* (see Giblett 1996, Chap. 5).

The Fens, and other wetlands, were not only bad for the body, but also bad for the soul, or for mental health, in the European philosophy and psychology of melancholy. Di Palma (2014, 19–22, 95) focusses her account of this aspect of wetlands on John Bunyan's *Pilgrim's Progress* in which as the scum and filth of sin is figured as 'the slough of despond' that would drag Christian down and engulf him for all eternity. The slough of despond is akin to the marsh of melancholy (as we saw in the previous chapter of the present volume). Like malaria, melancholy was associated with the miasmatic theory of illness. The miasmatic theory of melancholy arose with Robert Burton and John Bunyan in the seventeenth century, and went back through the Elizabethans of the sixteenth century, to the Greek philosophy of the four elements and humours (or psychosomatic states) that miasmatic gases caused melancholy (see Giblett 1996, Chap. 7). It went hand in hand with the miasmatic theory of malaria.

John Bunyan's *Pilgrim's Progress* was first published in 1678. Over half a century earlier in 1621, Robert Burton published *The Anatomy of Melancholy*. It was republished four times within the succeeding seventeen years. Bunyan in *Pilgrim's Progress* was not only singing from the song-sheet of *Psalm* 23: 4 in depicting the terrors of walking in 'the valley of the shadow of the death' (as Di Palma 2014, 20 notes), but also sermonizing in accordance with Burton's anatomizing of melancholy (as Di Palma does not note) in warning about the horrors of 'the slough of despond'. Miasma, malaria and melancholy are missing from *Wasteland*. They were part of the moralization of the Fens and other wetlands, and contributing factors in their drainage.

6.2 The Fens in Politics

Coming out of art history and lavishly illustrated with paintings, plans, maps and sketches, Di Palma's (2014) landscape history of wasteland devotes an entire chapter to 'Swamp,' or more precisely, to the Fens of England. It documents some appalling lowlights in the history of the English Fens, such as the class robbery perpetrated by their enclosure that went hand in hand with the ecocide, or more precisely the aquaterracide, of their draining (Di Palma 2014, 25, 35, 45, 52, 58, 66, 99–112; Ash 2016, 2017). In the middle of this conjunction and discussion, Di Palma invokes Karl Marx's scathing critique in Chaps. 27–31 of the first volume of *Capital* of the enclosure of 'the common land' (dryland and wetlands) into private property (Di Palma 2014, 55–56; Marx 1976, 877–926). Enclosure in short was what the late, great E. P. Thompson (1968, 237) called 'a plain enough case of class robbery.'

Marx argues that the enclosure of the commons and 'the clearing of the estates' (a euphemism for the dispossession and driving-off of the commoners, or local people) was the foundation for industrial capitalism and modern colonialism. In the final chapter (33) of *Capital I* entitled 'The Modern Theory of Colonization' Marx concludes that 'the expropriation of the mass of the people from the soil forms the basis of the capitalist mode of production' (Marx 1976, 934). It is a cruel irony of history that some Scottish settlers of western Victoria in Australia who had been subjected to 'the clearing of the estates' in Scotland later 'enacted their own brutal clearances on Indigenous [Aboriginal] peoples' (Wilkie cited by Giblett 2023, 128). Kiddle (cited by Giblett 2023, 88, 128) calculates that 'at least two-thirds of the pioneer settlers of the Western District [of Victoria] were Scottish. Nearly all were lowland farmers.'

A lack of sympathy or empathy from them towards Aboriginal people also did not extend to their acknowledging a similar clan culture in structure, even though some Scottish settlers of western Victoria recognized the similarities between the two societies (Wilkie cited by Giblett 2023, 128). This did not lead to the settlers expressing any fellow-feeling or exercising solidarity towards Aboriginal people. It was more a case of repetition and displacement, of repeating dispossession and clan antagonism against Aboriginal clans. The settlers were doing to others what had been done to them elsewhere and previously (rather than obeying Jesus' second great commandment of doing to others what you would have them do to you).

The aesthetic, political and economic were mutually reinforcing, as Raymond Williams (1973, 124–125) argued, to position and organize the land for profitable agricultural production and pleasurable aesthetic consumption. The intended consequence was to dispossess and drive off the traditional owners and users In the British enclosure movement 'from the second quarter of the eighteenth century to the first quarter of the nineteenth century,' as Williams (1973, 96) puts it, 'by nearly four thousand Acts, more than six million acres of land were appropriated' (though much of this land was specifically *wet*land, a crucial distinction[1]). Williams (1973, 96) goes on to point out that 'the process had been going on since at least the thirteenth century, and had reached a first peak in the fifteenth and sixteenth centuries.' The enclosure movement in wetlands was invariably associated with drainage to render the land profitable for monoculture with, as Christopher Hill (cited by Giblett 1996, 47–48) maintains, 'fen drainage alone extending the country's arable area by 10%.' In fact, the new outburst of drainage projects of the 1760s was contemporaneous with and supportive of the beginning of the agricultural revolution.

Marx's communist critique of the enclosure of common land into private property and of political economy by indicating that this was the basis for the capitalist mode of production is a salutary instance of the way in which environmental conservation is a working class and indigenous concern of local peoples preserving livelihoods in local places on local land in bioregional home-habitats by using earthly resources, not just a middle-class affectation of 'setting aside' special places in sanctuaries for their leisure, recreation and entertainment (for a critique of 'sanctuarism,' see Giblett 2011). Marx's work provides a platform for the 'socialist ecology' built upon later by Williams as a pioneer of the environmental humanities and the patron saint of eco-cultural studies in his trenchant critique of the enclosure of the commons, and in his work more generally on nature, the country and the city, landscape, livelihood and resources for a journey of hope (for further discussion, see Giblett 2020b, Chap. 10).

The expropriation of the soil from indigenous and local owners forms the basis of the capitalist mode of production and colonization in the English colonies where indigenous peoples' lands were taken by squatters (later legitimized after the fact) and granted freely to settlers and where they were the beneficiaries of this class robbery. At home and abroad, endo- and exo-colonization involved enclosure of the commons and 'clearing of the estates,' in the case of Australia, 'the biggest estate on earth' (Gammage 2011). What Marx (1976, 940) called 'the shameless squandering of uncultivated [*sic*] colonial land on aristocrats and capitalists by the English government' occurred not only in Australia (as Marx

[1] By ignoring this aspect of enclosure of the commons, Williams seems to have had a blind-spot about the specific history and geography of the Fens, even though he was living in them at the time (Williams 1973, 3), and even though he is the patron saint of eco-cultural studies (Giblett 2020a, Chap. 10). Perhaps he can be excused as he was following in the footsteps of Marx who does not mention the Fens either in his discussion of the enclosure of the commons into private property. All idols have feet of clay; all saints have blind-spots.

goes on to indicate), but also in Canada. Colonial land was not uncultivated, but had been cultivated by indigenous peoples for tens of thousands of years in the case of Australia and Aboriginal peoples (Gammage 2011; Pascoe 2014; Gammage and Pascoe 2021; Cumpston et al. 2022). Class expropriation of local and indigenous people and their lands (wet and dry), and class warfare against wetlands, has gone hand in hand with the draining, filling and destruction of wetlands ever since in England and in their colonial diaspora in Australia, Canada and the United States.

6.3 The Fens in Painting

Di Palma's *Wasteland* not only critiques the enclosure and draining of the Fens, but also presents some curious artefacts in the history of wetlandscape architecture, such as the design plans for the first artificial wetlands of 'duck decoys' to attract and 'harvest' water fowl by shooting them (Di Palma 2014, 115–119; see also Pryor 2019, 168–169[2]). *Wasteland* not only presents these appalling lowlights, but also showcases some appealing highlights in art history and landscape painting. English wetlands have not figured prominently in English landscape painting as 'the marshy Fens,' were, as Di Palma (2014, 123) puts it, 'the epitome of the unpleasing, unpicturesque landscape,' especially for such theorists and practitioners of the picturesque as William Gilpin. Wetlands were also neither sublime nor beautiful for such poets as William Wordsworth (see Giblett 2011, Chaps. 3 and 4).

For Di Palma (2014, 125–127) the epitome of the unpicturesque wetlandscape in paintings of the Fens are two gloomy depictions by John Sell Cotman (1782–1842). One is entitled 'Drainage Mills in the Fens, Croyland, Lincolnshire' (Cotman, *ca* 1835). The other is entitled 'St. Benet's Abbey' and is attributed to Cotman (*ca*1810). Both paintings are quite atypical of his entire body of work. In the catalogue for the

[2] Compare a similar and analogous contemporary situation in Oregon where a game 'hunter' constructed his own private, artificial wetland on the banks of the Columbia River and planted it with rice to attract waterfowl so he and his 'hunting' mates could take potshots at them and 'harvest' them as they fly through the east Pacific flyway (Giblett 2014, 3–4), a latter day instance of 'a duck decoy.' These two instances of wetlandscape architecture in America and England could be called industrial 'aquaorniculture' and contrasted with the constructing of wetland structures and the practicing of traditional aquaorniculture by Aboriginal people using nets to snare waterbirds (as we saw in the first chapter of the present volume).

exhibition held for the bicentenary of Cotman's birth in 1982, none out of over 130 works listed and discussed are of the Fens. Admittedly, 'the catalogue entries do not attempt to be exhaustive' (Rajnai 1982, 7). Indeed, none of the considerable holdings of Cotman's work in the Yale Center for British Art in New Haven, Connecticut, including the two paintings discussed by Di Palma, is included in this catalogue. The two paintings of the Fens in the Yale Center are also quite atypical of this collection and of the 275 holdings in the Cotman collection in the Yale Center for British Art (n.d.). Cotman is hardly the painter of the Fens.

A perhaps more likely contender for this accolade is Cotman's contemporary Peter de Wint (1784–1849) whose work was appreciated by John Clare (as we will see below) and recently by Pryor (2019, 391–392). De Wint painted two similar paintings both depicting the low-lying wetlandscape of the Fens as the background to a scene of heroic agricultural labour with the archetypal, if not clichéd, presence of a haywain mythologized by Constable as a symbol of taming the land and deriving benefit from it. One of these paintings is entitled 'Lincolnshire Cornfield near Horncastle' and is held in the Usher Gallery in Lincoln (de Wint 1815b); the other is entitled 'A Cornfield' and is held in the Victoria and Albert Museum in London (de Wint 1815a).

De Wint painted other 'low-lying landscapes in the Dutch manner' (Scrase 1979, 11), but none of these paintings are of the low-lying landscape of the Fens, a low-lying wetlandscape. They were primarily of rivers as he was 'partial to river scenery,' according to his wife Harriett (cited by Scrase 1979, 11). In other words, he was not partial to Fenland scenery, or perhaps more precisely, he was not partial to Fenland obscenery in the sense of what is hidden and secret. Besides river scenery, he was also partial to wifely scenery as he painted a stunning, erotically charged portrait of his wife with diaphanously veiled breasts, probably for his own private viewing pleasure and held in the Usher Gallery in Lincoln (de Wint n.d.). He did not paint landscapes with the same amount of passion, certainly not the Fens as they are largely absent from de Wint's entire oeuvre. They remain hidden and secret, unlike his wife's upper body. None of the one hundred drawings and watercolours listed and discussed in the catalogue for the exhibition held in the Fitzwilliam Museum in Cambridge in 1979 is of the Fens (Scrase 1979). De Wint did not depict pre-modern and mundane low-lying wetlandscapes, or when he did (such as in the two paintings mentioned) it was only as a background to agriculture.

6.4 The Fens in Poetry

The poet John Clare (1793–1864) was a contemporary of de Wint. Clare is for Pryor (2019, 77) 'the greatest poet of the English landscape.' He is also the greatest poet *for* the Fens and *against* enclosure. In Clare's poem, 'The Fallen Elm,' ruin was the guide of enclosure (cited by Pryor 2019, 79; Standing 2019, 14). Clare goes on to describe how 'The common heath became the spoilers prey' (Williams and Williams 1986, 85). In 'The Mores' and 'The Lament of Shady Well,' Clare goes on further to describe how 'vile enclosure' 'came and trampled on the grave/Of labours rights and left the poor a slave,' including Clare himself (Williams and Williams 1986, 91 and 98; cited by Standing 2019, 43). In 'Remembrances,' enclosure, like a Napoleon, 'let not a thing remain' (Williams and Williams 1986, 150). Enclosure was war against wetlands, the commons and their peoples, the commoners. For Clare, 'where enclosure has its birth/Its spreads a mildew o'er her mirth' (cited by Williams 1973, 136). Where the elm once stood are now 'workhouse prisons raised upon the site' (cited by Pryor 2019, 79). Enclosure, 'thou'rt a curse upon the land' (cited by Williams 1973, 136). Clare is specifically the greatest English poet against enclosure and for the Fens who wrote two long poems, 'The Fens' and 'To the Snipe' addressed to 'lover of swamps' (Williams and Williams 1986, 109–112, 150–153). In their commentary on Clare's poetry and prose, Williams and his daughter describe how 'the fens, as the poem of that name illustrates, are swarming with life'[3] (Williams and Williams 1986, 209). Clare is the patron saint of the Fens (Giblett 2020b, 34 and 140).

[3] 'Swarming' is a loaded biblical term as swarming creatures come under the Levitical interdiction of being abominable (*Leviticus* 11: 41–43; see also Douglas 1966, 56). In mainstream western culture wetlands have not only been seen as the habitat of swarming creatures, such as alligators and crocodiles who transition from earth to water, but also as abominable themselves as they mix the elements of earth and water. With the exception of this comment, Williams had nothing to say about Clare's poetry about the Fens in his and his daughter's critical commentary in this edition, including in their discussion of Clare as 'the finest naturalist in all English poetry' (Williams and Williams 1986, 213–217), earlier in their introduction to this edition (Williams and Williams 1986, 1–20) and in his otherwise exemplary discussion of Clare's life and work in *The Country and the City,* the pioneering work of ecocriticism and ecocultural studies (Williams 1973, 132–141). By downplaying this aspect of Clare's poetry, Williams seems to have had a blind-spot about the significance of the Fens for Clare, similar to his blind-spot about the history and geography of the Fens (see n1 above).

Clare praised de Wint for not indulging in 'modern fancy landscapes' (cited by Barrell 1972, 146). In a sonnet addressed to de Wint, Clare was more specific about what he meant by 'modern fancy landscapes' when he waxed lyrical that:

> [...] in thy landscapes I can well descry [...]
> No painted freaks no wild romantic sky
> No rocks or mountains as the rich sublime. (cited by Scrase 1979, 52)

Rather, what Clare found in de Wint's landscapes was 'the sunny truth/ Of nature' (cited by Scrase 1979, 52). He did not find the Fens depicted as the gloomy truth of nature (as he might have found in Cotman's paintings), nor did Clare find in de Wint's landscapes reeds and Fens as the fertile slime, nor 'swamps of wild rush-beds and sloughs' squashy traces' that he evoked in one of several short poems entitled 'Song' (cited by Barrell 1972, 148). Clare may have found the wild romantic sky-blue in the painted frock and the rolling hills—the pleasing prospect, as it were—of the breasts of de Wint's wife in his portrait of her, but he does not say. De Wint was not the painter of the Fens, however much Clare admired his work. Perhaps Nick Tearle, the self-proclaimed 'Fenland artist,' is the painter of the Fens (as we will see below).

6.5 The Fens in Prose

The English Fens have not figured prominently in English nature writing with the notable nineteenth-century exception of John Clare's poetry and the recent exceptions of Robert Macfarlane's essay, 'Ghost Species,' (2008) and Becky Owen-Fisher's poetry (Owen-Fisher and Tearle 2017). The Fens have figured more prominently in English prose fiction with Dorothy Sayers' *The Nine Tailors* (1934), Graham Swift's *Waterland* (2010b) and Stella Tillyard's *The Great Level* (2018) with all these novels being set in whole or part in the Fens.

Dorothy Sayers' 'great detective thriller' *The Nine Tailors* for Pryor (2019, 50) could 'only have been written by someone who knew the fen landscape and people intimately.' In fact, Sayers 'spent much of her childhood and early adulthood' in a Fenland village. It is hardly surprising then that Pryor (2019, 50) concludes that 'her affection for the place comes across vividly.' Early in *The Nine Tailors* Sayers (1934, 4) also paints a vivid, critical picture of Fenland drainage when she describes how 'the drain ran straight as a rule could make it, black and sullen, with a steep bank shelving down to its slow, unforgiving waters.' Similarly, Robert Macfarlane (2008, 109) recently remarked of the Fens how 'the landscape becomes rectilinear: ruler-straight road and field edges.' The straightness of the roads and dykes is the stock-in-trade description of the drained Fens in *The Nine Tailors* and 50 years later in Graham Swift's *Waterland* first published in 1983.

In depicting a scene reminiscent of Cotman, Sayers (1934, 5) goes on to describe how, 'at the end of a solitary mile, the gaunt shape of a windmill loomed up upon the farther bank of the drain, but no bridge led to it, and no light showed.' The windmill is imbued with menacing qualities, a monstrous emanation of a dark place. The gaunt windmill contrasts with the awe-inspiring church that Sayers (1934, 34–35) describes in much more intimate and loving detail than she does the Fens. As the daughter of a rector of the church in a Fenland village (Pryor 2019, 50), she was well-trained in architectural description, but also in theological exposition (as we will see later).

Of course, the Fens are not necessarily a pleasant place. One character 'with a faint London twang' describes them as 'not a nice part of the country' (65) and much later says 'it's a hell of a country' (324). Perhaps he is displaying the city-dwellers' prejudice against wetland country, city heaven versus wetland hell, and waste land. Sayers' (1934, 5 and 38) narrator evokes the Fens vividly in winter as 'this frozen desolation,' 'the flat, white wastes of the Fens, [with] the spear-straight, steel-dark dykes.'

By contrast, Nick Tearle's paintings, 'Winter at the Mille, West Deeping' and 'Inkerton Fen,' convey the flat whiteness of the Fens in winter, while his paintings, 'Ditch Near Etton,' 'King Street Drain' and 'Narrow Drain,' show the spear-straight, blue-light ditches and drains reflecting the sky (Tearle n.d.). The Fens for Tearle aren't wastes, though they are dark as in 'Rain on Morton Fen' and light as in 'Autumnal Sunset on Straight Drove,' 'Sunset over Six Score Road,' 'Fenland Morning,' 'Sunrise Near West Deeping,' 'Fenland Sunrise' and 'Sunset in the Fens,' and in-between as in 'Fenland Twilight' (Tearle n.d.). The Fens come out into the open and no longer remain hidden and secret in the paintings of Nick Tearle, the self-proclaimed 'Fenland Artist,' some of which accompany Becky Owen-Fisher's poetry (Owen-Fisher and Tearle 2017). Tearle conveys the changing light of different seasons and weather in the 'low flatlands' and 'flattened landscape' of the Fens that 'we call home' (Owen-Fisher and Tearle 2017, 3, 5).

Along similar lines to Tearle's paintings of the Fens in spring, such as 'April on Langtoft Fen' (Tearle n.d.), in *The Nine Tailors* Sayers' (1934, 69) narrator later evokes the Fens vividly in spring when, 'in its own limited, austere fashion, the Fen acknowledged the return of the sun. The floods withdrew from the pastures; the wheat lifted its pale green spears more sturdily from the black soil.' The Fens are fertile and well-watered in a wetting and drying seasonal cycle, the very qualities that make them so agriculturally productive. The Fens are no longer a dark, desolate, white wasteland, but a light, fecund, green wetland when 'the green fen spread out mile upon mile' (Sayers 1934, 74). Summer can be a different story, though, when 'that summer the water lay on the land [...] and the sodden ricks took fire and stank horribly' (369). With the eventual return of winter even the usually unflappable Wimsey remarked that 'they seem to keep a special brand of disgusting weather in these parts' (371).

Sayers has her most extensive discussion of the drainage of the Fens come from the mouth of her fictional detective, Lord Peter Wimsey, who describes some sections as 'rather more effectively drained' (63). Her most trenchant critique of drainage also comes from him as he describes some sections as 'bad engineering' (64). Sayers (1934, 197) saves her most prescient critique of drainage of the Fens until midway through *The Nine Tailors* when Wimsey remarks:

> if this country had been drained intelligently and all of piece [...] by running all the canals into the rivers instead of the rivers into the canals, so as to get a good scour of water, [the village] might still be a port and the landscape would look rather less like a crazy quilt. But what with seven hundred years of greed and graft and laziness, and perpetual quarrelling between one parish and the next, and the mistaken impression that what suits Holland must suit the Fens, the thing's a mess.

Mess indeed when land and water are out of place and in the wrong place. This is the kind of critique of drainage that the reader might expect to find more in a politically green environmental history of the Fens than in detective fiction.

The messiness of the Fens certainly comes to the fore in Sayers' (1934, 372) compelling description of a flood in the Fens, an invariable consequence of canalizing rivers, rather than vice versa:

> dark and menacing, the swollen flood-water raced through the sluices, eddying and turning and carrying the brown reeds and broken willow-stems and here and there fragments of timber filched from the drowned lands of the Upper Fen. And even while Wimsey watched, there came a change. Angry little waves and gurges ruffled the strong flow of the river, with an appearance as of repressed tumult and conflict.

The archaic 'gurges,' related to gorge as both ravine and throat, neatly combines a hydrological reference to drowning in a swirling eddy with an anatomical reference to swallowing in a monstrous act of oral sadism. Platt's (2019) account of 'the drowned world' across England and Wales in 'the great flood' of 2013–2014 follows in the muddy footsteps of Sayers' vivid description.

The flood is both a return of the repressed of the canalized water and a return to the repressed when the Fens were wetland. Sayers (1934, 373) goes on to describe:

> Dyke and drain were everywhere abrim and here and there the water stood in soaked fields as though they needed but little more to sink back into their ancient desolation of mere and fen.

Or should that be, 'merge back into their indigenous elation of mere and fen'? Just when Sayers seems to be going to make her most telling historical and ecological critique of drainage of the Fens, she retreats to a triumphalist chronology of ancient versus modern, waste land versus drained land.

Sayers (1934, 374) also retreats into a Christian theology of the wetland when she repeats the tried and true trope of the melancholy marshes: 'the air was so heavy with water' the sound of church bells 'drifted through the streaming rain with an aching and intolerable melancholy, like the noise of the bells of a drowned city pulsing up through the overwhelming sea.' On the second page of *The Nine Tailors* 'the sound of a church clock' prompts Wimsey to expostulate, 'thank God! Where there is a church, there is civilization' (Sayers 1934, 4). Where there is no church, as in the pre-Christian Fens, there is barbarism and paganism (or indigeneity and sacrality, I would prefer to say). Where there is a muted church, there is the melancholy marsh and the slough of despond. Where there is the muffled sound of church bells, there is a drowned civilization reminiscent of

Atlantis and Richard Jefferies' *After London* and prescient of J. G. Ballard's *The Drowned World* and George Turner's *The Sea and Summer.*

By contrast with Sayers, in Becky Owen-Fisher's poetry the sound of church bells is:

> Hitting the dull thud of fogged mystery,
> Calling, softly, straining against the wet darkness.
> They tell me to follow,
> Feet stumbling after their peal,
> Hands groping dank empty air [....]
> And the singing bells bring me through the bog [...]
> To the welcome booming peals. (Owen-Fisher and Tearle 2017, 12–13)

For Owen-Fisher, the sound of the bells in the Fens is a friendly guide in a foggy mystery, not a sign of civilization in a hell of a country, in a frozen waste land and in a melancholy marsh.

In *The Nine Tailors* the church is not only a sign of civilization, but also a refuge in time of flood as it stood on 'a little mound rising some ten or twelve feet above the level of the village, an elevation sufficient to save it in ancient times from inundation during the winter months' (Sayers 1934, 58). In time of flood in modern times, the church is an ark for the salvation of the Christian flock (animals had to fend for themselves) with the rector as latter day Noah. 'The drowned Fen' meant that 'the whole world was lost now in one vast sheet of water' (except for the ark/church) in a reprise of the biblical flood of gargantuan and global proportions. You can take the rector's daughter out of her father's church and rectory, send her to Oxford, commission her to translate Dante's *Divine Comedy* for Penguin classics, employ her in an advertising agency and get her to write Christian letters to a post-Christian world, but you can't take Christianity out of the detective writer. Like her fellow erstwhile Inklings in C. S. Lewis and J. R. R. Tolkien, Sayers' fiction was a vehicle for evangelizing Christianity. My *Swamp Deaths: Collected Cold Cases and Other Marshy Mysteries* (Giblett 2022) is a vehicle for promoting pro-paludal wetland culture. The result, or down-flow of the flood, is that 'the Fen had reclaimed its own' (Sayers 1934, 390 and 391). Or more precisely, the water of the drained and reclaimed Fens had re-reclaimed the Fens. The result is the return of, and to, the repressed wetland with all its glorious, messy, moralized and odiferous features of 'the dismal strand of ooze and weed' (Sayers 1934, 395).

Fifty years later Graham Swift's *Waterland* (2010b), 'his great novel of Fenland life' for Edward Platt (2019, 133), was published in 1983 to critical acclaim and commercial success culminating in the film of the book starring Jeremy Irons. Crick, the first-person narrator, acknowledges like Sayers, the smell of the Fens, 'which is smelt over and over again in the Fens. A cool, slimy but strangely poignant and nostalgic smell' (11). Sliminess is a dominant feature of the Fens, as is flatness. Crick, like Sayers' narrator, offhandedly relates now 'the land in that part of the world is flat. Flat, with an unrelieved and monotonous flatness,' its 'uniform levelness broken only by the furrowed and dead-straight lines of ditches and drains' (10). The Fenlanders' answer to the question, 'why are the Fens flat? is typically 'so God has a clear view' (21). It 'a landscape which, of all landscapes most approximates to nothing' (21). In this 'nothing-landscape [...] Fenlanders do not try to hide—since we know God is watching' (58).

Similarly, Owen-Fisher emphasizes the flatness of the Fens. In a poem entitled, 'The Flatland,' Owen-Fisher (2017, 17) describes 'saturated ditches' as 'Brimming with sleek unreal creatures,/ A waterworld of history, bubbling under.' This history bubbles up from below in the history of enclosure and conservation conveyed in the poem, 'The Green Man.' Owen-Fisher (2017, 20) writes how:

> He loves that land, the flat, never ending stretch to the sea.
> He can picture it in his mind's eye, the wide expanse of nothing.
> And everything.
> Full of the feeling of home.

The creatures and trees 'hold his muddy green heart inside' and 'the things that grow in his flattened homeland, take root inside him.' The green man is part and parcel of the Fens, and they are inside him. They are not a landscape; they are nothing and everything to him.

By contrast, in *Waterland* the flatness of the Fens is a business opportunity that makes 'this nothing-landscape' ideal as 'a drawing-board for the plans' of prosperous maltsters (Swift 2010b, 25). The Fens are a flat surface on which to draw the straight lines of their plans for improvement and a screen on which to project their phantasies of prosperity. Reclamation renders the sometime turbulent surface of water into the flat surface of land. Crick remarks that 'the chief fact about the Fens is that they are reclaimed [from what? for what? for whom? one wonders], land that was once water, and which, even today, is not quite solid' (Swift 2010b, 16). Like Sayers' narrator, Crick wonders of land reclamation, 'is it desirable, in the first place, that land should be reclaimed?' (17). The desirability of drainage is questionable indeed; in all probability and with the benefit of hindsight, drainage was undesirable. For the non-Christian Crick, unlike the Christian Sayers, the Fens are also 'a fairy tale land' and 'a magical, a miraculous land' (11 and 121). Indeed, 'the bare and empty Fens yield so readily to the imaginary—and the supernatural' (25) with 'marsh-sprites and will-o-the-wisps' (26).

Unlike Di Palma, Swift is attentive to other pertinent aspects of the Fens, such as the theory of the humours based on the philosophy of the four elements, earth, air, fire and water. The humours mix the elements with the mixing of water and earth producing the humour of phlegm (as we saw in a previous chapter of the present volume). Fenlanders are phlegmatic people with a phlegmatic humour and a phlegmatic sense of humour. It goes with the place as if they have absorbed some of its characteristics. Crick describes how 'phlegm [is] a muddy, silty humor' (Swift 2010b, 22) as 'the Fens were formed by silt [...] which demolishes as it builds' (16). Phlegm, 'or slime,' is 'an ambiguous substance. Neither liquid nor solid: a viscous semi-fluid [...] It deters the sanguine and the choleric and inclines towards melancholy' (341). Indeed, 'melancholia and self-murder are not unknown in the Fens' (24), and 'poor Fenland madmen and melancholics' (100) are not uncommon too. The mixing of water and earth also produces melancholy.

The phlegmatic is not stable in the periodic table of elements and humours. Even though the melancholic has traditionally been associated with the cold and dry, a visit to the cold and wet/dry marshes by the drylander can induce melancholy, whereas the indigenous wetlander by temperament is usually phlegmatic, but tends towards the melancholic. By temperament, the wetlander at home in the wetland is phlegmatic (associated with the watery), whereas the drylander visiting the wetland tends to be infected, or more precisely affected, by the melancholic (the earthy, though more precisely the earthy in combination with the watery).

In 2008, Swift (2010a, vi) wrote an introduction for the 25th anniversary edition of *Waterland*. In it he acknowledges that initially in writing the novel 'the Fenland setting, which could hardly seem more prosaic and monotonous' (even unpromising) was 'a special, almost literal stage' on which the action and the lives of the character are played out and the arc of the story takes place. The Fens in the beginning of writing the novel 'may have seemed to me the ideal non-setting, the ideal flat, bare platform for my human drama' (vii), as if the Fens were the mere passive stage on which, and backdrop against which, the action of Swift's human drama took place. Swift goes on to relate that 'what quickly happened [in the writing process] was that the apparent background became a foreground, *even a kind of principal character*' in the novel (vii; my emphasis). Wetlands, like the Fens, have agency, like characters in a human drama, like humans in the drama of real life.

Novels are invariably pre-occupied, if not obsessed, with human drama played out against the backdrop of a setting for the action. They do not usually have a principal character who is not human, certainly not a wetland, such as the Fens, that figure in the non-human drama, the drama of nature and the elements. Novels can have a non-human principal character in the form of an animal, as in the famous case of the white whale Moby Dick in Herman Melville's novel of the same name, for many the greatest novel in English for the very reason it is not only about Captain Ahab's obsessive quest, but also about sea-faring, whales, whaling, whiteness and nature.

Besides the Fens themselves, *Waterland* has another non-human principal character in the form of eels 'who feature prominently in the book' as Swift (2010a, v) says in his introduction with a chapter in the novel devoted to their natural and cultural history (Swift 2010b, 197–206). Having a habitat and wetlandscape, such as the Fens, as a principal character is another matter entirely, at least until *Waterland* came along. Following in Swift's footsteps, Liam Davison's 1993 novel *Soundings*, a kind of Australian *Waterland*, set in the Koo-Wee-Rup or Great swamp of south-eastern Australia outside Melbourne, and Peter Matthiessen's 1990s trilogy set in the Florida Everglades beginning with *Killing Mister Watson* are perhaps other cases in point of novels with a wetland as a principal character (as we have seen in previous chapters of the present volume).

In his 2008 introduction, Swift (2010a, vi) also acknowledges that his 'physical research for the novel was minimal.' And it shows. He did not seem to have spent much time walking, punting or sitting in the Fens absorbing local sounds, colours and smells and writing about them in a journal as the basis and sourcebook for the novel, unlike Macfarlane's (2008) 'Ghost Species,' a recent instance of 'new nature writing' about the Fens. *Waterland* is certainly not a book of nature writing about the Fens. Swift is not the Thoreau of the Fens. *Waterland* is not to the Fens as

Thoreau's *Walden* is to Walden Pond and Woods in which they are not only principal characters, but also they, and their plants and animals, are minutely and lovingly described from first-hand observation and experience. Yet it is the physical aspect of the novel that meant Swift (2010a, vi) had to disabuse his Fenland readers that he is not a fellow 'Fenny.'

Although his physical research for the novel was minimal, in his 2008 introduction Swift (2010a, b, vii) also goes on to acknowledge, 'I did quite a lot of historical research,' including, but not acknowledging, Charles Kingsley's (1884) 'prose idyll,' 'The Fens.' Swift's historical research shows too, as does Macfarlane's in his 'Ghost Species.' Macfarlane (2008, 111) also did some geographical and geological research as he describes the Fens as an 'East Anglian [...] low-lying area of around 1,200 square miles,' amounting to 'one million acres' as Pryor (2019, xii) points out. Macfarlane then goes on to discuss the geology of the Fens with its surrounding limestone hills, chalk and sand. Geology is literally the base for history and geography.

Like Di Palma (2014, 101–102), other writers, such as Swift (2010b, 18–19), Macfarlane (2008, 111–112), Pullman (2011, 99), Ash (2016, 2017), Platt (2019, 137–138 and 141–143) and Pryor (2019, xv, 326–332), trace the history of the drainage of the Fens engineered by Cornelius Vermuyden and others in the seventeenth century. *Northern Lights*, the first part of Pullman's trilogy *His Dark Materials*, describes the Fens graphically as:

> that wide and never fully mapped wilderness of huge skies and endless marshland in eastern Anglia. The furthest fringe of it mingled indistinguishably with the creeks and tidal inlets of the shallow sea, and the other side of the sea mingled indistinguishably with Holland; and parts of the Fens had been drained and dyked by Hollanders [...] But parts of it had never been drained or planted or settled at all, and in the wildest central region [...] eels slithered and waterbirds flocked, [...] eerie marsh-fires flickered and waylurkers tempted careless travellers to their doom in the swamps and bogs [...] (Pullman 2011, 99)

No reasons are given for dyking and draining, and no consequences are drawn. They seem to be merely facts of history to note and pass over without comment.

Enclosure of the commons into private property produced highly fertile and arable land for farming as we have seen and as Macfarlane (2008,

112) notes when he describes how the draining of the Fens 'revealed hundreds of thousands of acres of the most voluptuously rich soil in England'—enough to make farmers salivate and lust after it. Pryor (2019, xv and 342) concurs that, 'once drained, fenland soils are very fertile,' even 'exceptionally fertile.' Swift notes the unintended consequence that 'reclaimed land shrinks' (Swift 2010b, 19). This history is the context 35 years later for Stella Tillyard's historical romance *The Great Level* (2018) that seems indebted also to Charles Kingsley's (1884) 'prose idyll,' 'The Fens,' that gives a wonderful potted history of the drainage of the Fens.

Tillyard's first-person narrator, Jan Brunt, is a Dutch engineer who is employed by Vermuyden in the drainage of the Fens in the seventeenth century that began in the reign of Charles I and was completed in the 'Commonwealth' of Oliver Cromwell. Royals and republicans had a similar antipathy to wetlands and shared a capitalist drive to 'improve' them by draining, converting them into dryland agriculture and deriving profit from them. The prologue to the novel is set on Manhattan Island in Dutch New Amsterdam (later New York) where Brunt has ended up and where he still pursues obsessively the Dutch compulsion 'to attempt the separation of land and water wherever we find them muddled up' (Tillyard 2018, 7), and muddied together, one might add. The means to separate land and water from each other in what Macfarlane (2008, 112) calls the 'terraqueous' (literally 'earth-water') Fens is by what Brunt calls 'embanking' whose aim is 'to make an edge, a clear boundary and separation' between land and water (Tillyard 2018, 8). 'Embanking' is his preferred euphemism for draining and reclaiming marshes and meres (see 96–97, 131–132 and 140). He prides himself that under his supervision 'where once the land slid under the water in a disorderly way, the two are now divided by pilings that have given [the island] a firm edge all round its [southern] tip' colonized by the Dutch (8; see also p.43).

The Great Level reminds the reader that Manhattan Island was Manahatta, an island of marshes, and that New York is a city of ghost marshes (see Giblett 2016, Chap. 9). Macfarlane (2008, 123) suggests that the Fens are a place of 'ghost species,' 'a species that had been out-evolved by its environment, such that, while it continues to exist, it has little prospect of avoiding extinction. Ghosts endure only in what conservation scientists call "non-viable populations".' This is a sad and sorry story of species' loss. The harsh terms of so-called conservation science belie the heart-ache and sense of loss experienced by Fenlanders. The ghost swamps and marshes of the Fens, and of the cities built on or in them, endure in what Walter Benjamin (1999, 462, 464) called a dialectical image in which 'what has been within a particular epoch is always, simultaneously, what has been from time immemorial.' 'The primal phenomenon of history' of the swamp world is 'present by virtue of its very oblivion.'

The novel proper begins with Brunt meeting Vermuyden and being taken into his employ as an engineer in the drainage of the Fens. Vermuyden describes them as 'a great wilderness [...] lying unimproved' (Tillyard 2018, 21). Brunt asks, 'is it a lived wilderness, or at present water only?' to which Vermuyden replies, 'it is inhabited, yes, Jan; but, the island of Ely excepted, by a lazy and barbarous people who trap eels and other such trash foods.' Ironically, 'Ely' means 'island of the eels' (Macfarlane 2008, 112) so the name of the good Christian island is named after the trash food of the local barbarians! Brunt persists, 'does not the place belong in some manner to these people, sir?' to which Vermuyden replies, 'indeed not, or only by custom, for much rightly belongs, as underwater, to the realm, having formerly been the property of the King' (Tillyard 2018, 21–23). Customary belonging is the common law, the law of the commons. The custom of the commons meant the Fens belonged in common to all and were shared by all. This was the realm before it became crown land and before it was enclosed into private property. Vermuyden scoots over this history and legitimates 'improvement' (and Pryor (2019, 342) follows suit) by way of drainage, embankment, enclosure of the commons and dispossession of Fenlanders. *The Great Level* traces in part this process, though it does not mention the commons, nor does Pryor (2019, 342).

Vermuyden's denigration of the Fens as being inhabited by barbarous people is reinforced by one tavern-goer who tells Brunt he has 'heard it said that barbarous people live there.' Another one adds 'barbarous and godless, too, so I was told' (Tillyard 2018, 27). This is all second-hand hearsay that *The Great Level* contests as the Fenlanders are barbarous 'it is said' (35), but they call all others uplanders, so the Fenlanders are 'lowlanders,' like the Dutch. Brunt relates later that 'witches are said to be common amongst them, and many, I heard an old man say, still practise the pagan religion that long ago held sway across these islands' (49). In other words, the pagan religion is implicitly pre-Christian and explicitly barbarous, the religion of a barbarous people in a barbarous place. It is implicitly prior and opposed to Christian civilization and its churches as presented in *The Nine Tailors*. The Fens are also, like *Waterland*, a place of the supernatural with 'marsh lights [and] will-o-the-wisps conjured by spirits' (119).

Brunt travels from Cambridge to Ely and is 'heartened by the flatness of the country' as he travels 'across a flat and marshy plain' and knows he is 'going into the fenland' (30). He has 'reached a world where water has its empire still' (31). Similarly, for Macfarlane (2008, 109) 'entering the Fens always feels like crossing a border into another world.' What kind of world the Fens are Macfarlane does not specify other than pointing out different species of trees and sedges and implying the Fens are the world of the other, a wet world to the dry world, of a wetland people to dry-land people.

For the Fenlanders in *The Great Level* the Fens are an expression of 'the watery underworld' (Tillyard 2018, 122) with no pejorative overtones attaching to the concept of the underworld. The Fens are not hell, but heaven, albeit a watery world with watery human inhabitants who are described as amphibious for Macfarlane (2008, 111). This tends to make them sound like they were frogs, and indeed this was a pejorative jibe directed at them, but they were perfectly adapted to their watery habitat and home.

Brunt has come to wrest the watery world of the Fens from the empire of water by draining 'the drowned country' and 'the green world' (Tillyard 2018, 31) and thus making the wet land into dry land and so 'redeem the drowned land' (54; see also p.91). Drainage of wetlands and Christian redemption of the pagan went hand in hand. Indeed, for the Royalist William Dugdale (cited by Platt 2019, 15) in the seventeenth century 'works of Drayning are most antient and of divine institution.' Platt (2019, 15) comments that in 'Christian mythology' (or more precisely in Dugdale's version) God is 'a kind of universal drainage engineer' who reverses the work of God as the universal wetland creator and who legitimates human drainage engineering. Watery land can be drained, but 'the watery air' (Tillyard 2018, 31) cannot be drained or dried. It proves a problem as a health hazard under the miasmatic theory of malaria, literally 'bad air,' or ague that 'the meres breed' from their 'miasmic air' and that 'haunt the ground' like malevolent ghosts' (4, 102, 117, 172, 180 and 186; see also Pryor 2019, 4–5).

The flatness and wateriness of the Fens prompts Brunt to note that when the creator 'created the earth, it was without form and covered all over with water, just as this land is in winter' (Tillyard 2018, 53). He is alluding to *Genesis* 1: 2 in the Authorized/King James version of the Bible first published in 1611 before the events of *The Great Level.* The creator then divided the land from the water on the second day, as Brunt also notes (53). Brunt later draws the implication that 'water was God's great love, the first thing, the dry land only a second thought' (76). Water covering the earth is the creator's (now named God) first and best work on the first day of creation. It was all downhill after that. Brunt does not draw the implication that drainage is fallen 'Man's' last and worst work of doing God's work for 'Him,' and becoming God-like, arrogating to himself the position and function of God, like building a city to heaven (as in *Genesis* 11: 4). Wetlands are places that God forgot when 'He' divided land and water. They hark back to his first and best work. After the Fall from grace, arrogant 'Man' divides land and water from each other (and builds cities to the heavens in drained wetlands).

Like Swift's *Waterland, The Great Level* acknowledges the theory of the elements (and the humours in passing (192)) and the silty nature of the Fens (as we saw with *Waterland* in a previous chapter of the present volume). Brunt describes how, 'in the marshes, where there might be a boundary [between land and water], but is none, the two elements are fused. The land is water-soaked and the water carries its heavy load of silt' (73). The job of drainage is to defuse the two elements of land and water by dewatering the land. Brunt goes on to stated that water 'must be drawn out from the land [...] and insinuated into a clean division of liquid and solid' (73). 'Insinuated' implies something covert, surreptitious and underhanded as the dictionary confirms: 'slide (oneself or a thing) slowly and smoothly into a particular place; manoeuvre oneself into (a favourable position) by subtle manipulation.' Just as the land slid under water in the marshes of Manhattan and the Fens, so drainage slides the land out from under the water and so divides the land and water, solid and liquid. It brings the watery underworld out from below into the earthy over-world.

Brunt realizes that 'for one world to be made, another must die' (99). For the new world of dryland to be made, the old world of wetland must die in the capitalist logic of drainage engineering and private property. He has seen 'that this unimproved world has its own way of being which will be lost' (99). For him this is a fact of life and of history, not a loss to be mourned. He is unlike his lover and protégé, the Fenlander Eliza (fenny, eely Eliza versus urban island Ely; barbarous, pagan Fens versus Christian church and civilization). She drains the swamp on her farm in Virginia (a long, romantic story). In the process, she has 'to fell the cypresses which groaned as they came down into the swamps' and she is 'pained by the loss' (209). She imbues the trees with sentience.

In *The Great Level* Eliza is a mediating figure between the ancient world of the Fens and the modern world of draining and reclaiming wetlands, such as the Fens of eastern England, the marshes of Manhattan and the swamps of Virginia. She has a foot in both worlds and tries to have the best of both, though in many ways they are antithetical to each other as a drained and undrained wetland is not possible. The Fens become the template for Brunt's 'embanking' of Manhattan marshes and Eliza's draining of Virginian swamps. Yet whereas Brunt transfers the template unerringly, Eliza modifies it in more environmentally sensitive ways as she has 'maintained several swamps away from the immediate surroundings of the house' (209). She wants to have her cake and eat it too.

The unimproved world of the wetland is its own world, complete and sufficient to itself. It is a way of being, an indigenous sacrality (see Webb 1996). It is not a social construction, nor a discourse, an institutionalized way of seeing, saying and doing, such as in Vermuyden's 'Discourse Touching the Draining of the Great Fennes.' The Fenlanders say 'we have always been here' (121). They are people of the place. The Fenlanders belong to the Fens and they belong to them by 'the right of custom' (121) in the customary law of the commons prior to and without a document of written law. The people belong to the land and the land belongs to the people, as in the Aboriginal custom and concept of 'Country' (see also Webb 1996).

The sacrality of the Fens is encapsulated by two figurines that are introduced in the 'Prologue' when Brunt is in New Amsterdam (12) and that were given to him earlier in the Fens as related later (137). He carries them around in his pocket, polishes them, sits them on his desk during the day while he is writing and locks them up at night (137). They are antiquities that he has collected and displays (I am reminded of Freud and his desk) and fetishes that he worships as sacred objects. He relates that 'they are women, goddesses or witches […] They are old women, many times mothers' (137). They are the Great Goddess, or Great Mother, of the swamp world (as discussed by J. J. Bachofen and catalogued by Marija Gimbutas; see Giblett 1996, 145–50, 196–198).

One figure is baked from clay, seeming to 'symbolize' or convey the element of the earth, the other carved from sparkling rock, symbolizing or conveying the element of water. In the clay figure 'the gash of her belly button looks like an opening to the underworld' (12); indeed, it *is* an opening to the underworld of her swampy womb, and to the underworld of the womby swamp. Or more precisely and less coyishly, 'her navel and the secret place between her legs' are described as 'black as caves' (137). The slimy womb and womby wetland has black water, an uncanny abject of fascination and horror.[4]

By contrast, 'the other woman is fish-like and liquid [...] Legs fused like a mermaid's tail [...] She is a sea creature' (13). Brunt understands that 'they belong together' (p137). They belong together as earth and water belong together in the marshy and swampy world and womb of the wetlands of the Fens. Separating them out completely is misguided, as 'without water, land is lifeless and barren. Wish then for the two together' (188). The womby wetland dries up when it is drained and cannot procreate new life. Land and water belong together, as they did in the Fens before 'embanking' and as they did on the first day of creation where they were God's first and best work. Long live the Fens!

References

Ash, Eric H. 2016. Reclaiming a New World: Fen Drainage, Improvement, and Projectors in Seventeenth-Century England. *Early Science and Medicine* 21: 445–469.

———. 2017. *The Draining of the Fens: Projectors, Popular Politics, and State Building in Early Modern England.* Baltimore, MD: Johns Hopkins University Press.

Barrell, John. 1972. *The Idea of Landscape and the Sense of Place 1730–1840: An Approach to the Poetry of John Clare.* Cambridge: Cambridge University Press.

Benjamin, Walter. 1999. *The Arcades Project.* Trans. H. Eiland. Cambridge, MA: Belknap Press of Harvard University Press.

Cotman, John Sell. ca 1835. Drainage Mills in the Fens, Croyland, Lincolnshire. https://collections.britishart.yale.edu/vufind/Record/1666774. Accessed 7 Jan 2020.

———. Attrib. "St. Benet's Abbey". https://collections.britishart.yale.edu/vufind/Record/1666197. Accessed 7 Jan 2020

Cumpston, Zena, Michael-Shawn Fletcher, and Lesley Head. 2022. *Plants: Past, Present and Future.* Port Melbourne: Thames & Hudson.

de Wint, Peter. 1815a. A Cornfield. http://collections.vam.ac.uk/item/O15129/a-cornfield-oil-painting-peter-de-wint/. Accessed 7 Jan 2020.

[4] For further discussion of the uncanny drawing on Sigmund Freud's work, see Giblett (1996, 31–34), and on his work and others, such as Friedrich Schelling, see Giblett (2019).

———. 1815b. Lincolnshire Cornfield Near Horncastle. https://artuk.org/discover/artworks/lincolnshire-cornfield-near-horncastle-81680/view_as/grid/search/keyword:lincolnshire-landscape%2D%2Dactor:de-wint-peter-17841849/page/1. Accessed 7 Jan 2020.

———. n.d. Mrs Peter de Wint. https://www.artuk.org/discover/artworks/mrs-peter-de-wint-17911866-81833. Accessed 7 Jan 2020.

Di Palma, Vittoria. 2014. *Wasteland: A History*. New Haven, CT: Yale University Press.

Douglas, Mary. 1966. *Purity and Danger: An Analysis of the Concepts of Danger and Taboo*. London: Routledge.

Gammage, Bill. 2011. *The Biggest Estate on Earth: How Aborigines Made Australia*. Sydney: Allen & Unwin.

Gammage, Bill, and Bruce Pascoe. 2021. *Country: Future Fire, Future Farming*. Port Melbourne: Thames & Hudson.

Giblett, Rod. 1996. *Postmodern Wetlands: Culture, History, Ecology*. Edinburgh: Edinburgh University Press.

———. 2011. *People and Places of Nature and Culture*. Bristol: Intellect Books.

———. 2014. *Canadian Wetlands: Places and People*. Bristol: Intellect Books.

———. 2016. *Cities and Wetlands: The Return of the Repressed in Nature and Culture*. London: Bloomsbury.

———. 2019. *Environmental Humanities and the Uncanny: Ecoculture, Literature and Religion*. London: Routledge.

———. 2020a. *New Lives of the Saints: Twelve Environmental Apostles*. Lanham, MD: Hamilton Books.

———. 2020b. *Odds and Ends of a Writer's Life 2014–2019*. Champaign, IL: Common Ground.

———. 2022. *Swamp Deaths: Collected Cold Cases and Other Marshy Mysteries*. London: Europe Books.

———. 2023. *Middlemarsh: The Hopkins River, Kindred Wetlands and Remarkable People*. London: Transnational Press.

Hillsdon, Mark. 2020. Dry Run: The Wet Farming Experiment that could Sow Seeds for Future Crops. *The Guardian*, November 12. https://www.theguardian.com/environment/2020/nov/12/dry-run-the-wet-farming-experiment-that-could-sow-seeds-for-future-crops-aoe. Accessed 13 Nov 2020.

Kingsley, Charles. 1884. The Fens. In *The Works, Volume XV, Prose Idylls*, 93–131. London: Macmillan.

Macfarlane, Robert. 2008. Ghost Species. In *Granta: The Magazine of New Writing 102: The New Nature Writing*, 109–128.

Marx, Karl. 1976. *Capital: A Critique of Political Economy, Volume One*. Trans. B. Fowkes. London: Penguin Books in association with *New Left Review*. First published in German in 1887.

Owen-Fisher, Becky, and Nick Tearle. 2017. *Standing High out of Shrunken Peat.* N.p.: Compass Publishing.

Pascoe, Bruce. 2014. *Dark Emu: Black Seeds: Agriculture or Accident?* Broome: Magabala.

Platt, Edward. 2019. *The Great Flood: Travels Through a Sodden Landscape.* London: Picador.

Pryor, Francis. 2019. *The Fens: Discovering England's Ancient Depths.* London: Head of Zeus.

Pullman, Phillip. 2011. *His Dark Materials: Northern Lights, the Subtle Knife, the Amber Spyglass.* London: Everyman's Library.

Rajnai, Miklos, ed. 1982. *John Sell Cotman 1782–1842.* London: Herbert Press.

Sayers, Dorothy. 1934. *The Nine Tailors.* New York: Harcourt.

Scrase, David. 1979. *Drawings & Watercolours by Peter de Wint: A Loan Exhibition Inaugurated at the Fitzwilliam Museum, Cambridge.* Cambridge: Cambridge University Press.

Standing, Guy. 2019. *The Plunder of the Commons: A Manifesto for Sharing Public Wealth.* London: Penguin.

Swift, Graham. 2010a. Introduction to the 25th Anniversary Edition. In *Waterland: With a New Introduction by the Author*, v–xi. London: Picador.

———. 2010b. *Waterland: With a New Introduction by the Author.* London: Picador.

Tearle, Nick. n.d.. Fenland Paintings. https://nicktearle.net/fenland-paintings. Accessed 7 Jan 2020.

Thompson, E.P. 1968. *The Making of the English Working Class.* Harmondsworth: Penguin.

Tillyard, Stella. 2018. *The Great Level.* London: Chatto & Windus.

Webb, Hugh. 1996. 'Aboriginal Country: Not a Construction, A Way of Being.' In *Western Australian Wetlands: The Kimberley and South-West*, edited by Rod Giblett and Hugh Webb, 61–76. Perth, Western Australia: Black Swan Press/Wetlands Conservation Society.

Williams, Raymond. 1973. *The Country and the City.* London: Chatto & Windus.

Williams, Merryn, and Raymond Williams, eds. 1986. *John Clare: Selected Poetry and Prose.* London: Methuen.

Yale Center for British Art. n.d. Cotman. https://collections.britishart.yale.edu/?utf8=✓&join=AND&lookfor0%5B%5D=&type0%5B%5D=allfields&bool0%5B%5D=AND&lookfor1%5B%5D=&type1%5B%5D=title&bool1%5B%5D=AND&lookfor2%5B%5D=Cotman&type2%5B%5D=auth_&bool2%5B%5D=AND&lookfor3%5B%5D=&type3%5B%5D=earliestDate&bool3%5B%5D=AND&lookfor4%5B%5D=&type4%5B%5D=topic&bool4%5B%5D=AND&lookfor5%5B%5D=&type5%5B%5D=geographic&bool5%5B%5D=AND&search_field=all_fields&q=Cotman. Accessed 7 Jan 2020.

CHAPTER 7

Bog Irish

7.1 Seamus Heaney: Patron Saint of Bogs

Seamus Heaney was born on 13 April 1939 in Ireland and died there in 2013. Besides winning the Nobel Prize for literature in 1995, he has been canonized as St Seamus of Ireland, the patron saint of bogs. He converted the Irish from Roman Catholicism to Irish bogism after St Patrick had converted the bog Irish to Roman Catholicism. St Seamus also reintroduced bog monsters to Ireland after St Patrick had driven them out. Supposedly, St Patrick drove out snakes from Ireland, but what he really drove out were monstrous bog serpents, or dragons, because they were 'pagan' and because he was converting the bog Irish to bogless Christianity. What is Ireland without bogs? Just another benighted and bogless Christian country.

The life of present Irish bogs and the death of absent Irish bogs are commemorated and celebrated on St Seamus' Day, April 13, his birthday. Sometimes April 13 falls on a Friday. When this occurs, St Seamus' Day is colloquially called 'Black Bog Day,' the antidote and cure-all for the 'black dog' of depression and melancholia. St Seamus' Day is the annual occasion for bog-racing, which involves half-swimming and half-slogging through Irish bogs carrying one's partner across one's shoulder.

The sacred texts or bible of Irish Bogism are Seamus Heaney's poems, or at least they are the New Testament. The old testament of Irish Bogism

R. Giblett, *Wetland Cultures*, Literatures, Cultures, and the Environment, https://doi.org/10.1007/978-3-031-57365-1_7

is Lord Dunsany *The Curse of the Wise-Woman* first published in 1933, with the apocrypha by the American H. P. Lovecraft in his weird story, 'The Moon Bog,' with all three bodies of work comprising the bible of Irish Bogism. In the present century, Kelly Barnhill in *The Girl who Drank the Moon* published in 2017 sings the song of the bog. A supplement of 'Rebirth in a Bog' complements this eminent canon of the bog bible in the conclusion to the present chapter. It is excerpted from my book of detective stories, *Swamp Deaths* (Giblett 2022). The present chapter traces the ancient, traditional and contemporary culture of Irish bogs in twentieth- and twenty-first century poetry and prose.

From the earliest of his published poetry, Seamus Heaney celebrated the centrality of the bog to the Irish landscape, to Irish history and to Irish cultural identity. He even went so far as to suggest in 'The Tollund Man,' one of several poems devoted to 'bog men,' that 'I could risk blasphemy,/ Consecrate the cauldron bog/Our holy ground [...]' (Heaney 1980, 126). Heaney sees the wetland as a sacred place in a conservation counter-theology. Part of the holiness of the bog consists in its distinctive Irishness, particularly to the Irish landscape. Heaney seeks and finds a sense of Irish cultural, if not national, identity in the bog, as does his antecedent in Lord Dunsany (2014, 132) for whose first-person narrator in *The Curse of the Wise-Woman* the bog seems to be 'the very heart of Ireland.' For the wise-woman herself there is no 'seems' about it as she says, '[i]t's the heart of Ireland' (165).

The bog has been associated with death, and with life. Heaney feels kinship with the bog, or at least he feels 'kinned by hieroglyphic/peat' which writes on him as the swamp oak wrote on Thoreau's heart. In the poem 'Kinship' he writes:

> Quagmire, swampland, morass:
> the slime kingdoms,
> domains of the cold-blooded,
> of mud pads and dirtied eggs.
>
> But *bog*
> meaning soft,
> the fall of windless rain,
> pupil of amber.
>
> Ruminant ground,
> digestion of mollusc

and seed-pod,
deep pollen bin.

Earth-pantry, bone-vault,
sun-bank, embalmer
of votive goods
and sabred fugitives.

Insatiable bride.
Sword-swallower,
casket, midden,
floe of history.

Ground that will slip
its dark side,
nesting ground,
outback of my mind. (Heaney 1980, 196–197)

By contrast to Thoreau for whom there is a hard bottom everywhere, for Heaney 'the wet centre is bottomless.' By contrast also to his earlier compatriot poet, W. B. Yeats, for whom 'the centre cannot hold,' for Heaney 'this centre holds' despite, or perhaps because of, its fluidity (Heaney 1980, 86 and 198). Thoreau was looking for some sort of bed-rock of reality, whereas for Heaney there is none; Yeats was looking for some fixed reference point in the flux of modernity, whereas for Heaney the only centre which holds in the mire of hypermodernity is one that changes like the bog.

Heaney is fascinated by what he calls 'a mothering smell of wet,' though perhaps not emanating from a bog (Heaney 1979, 53). The bog is 'a slimy birth-cord' which would attach him too strongly to the spawn of mother earth, or specifically to the 'Bog Queen' of Ireland herself, or which would drag him down into 'the black maw/Of the peat,' into its orally sadistic vagina dentate (Heaney 1980, 186 and 189). After being immured in the bog and masticated by peat's teeth, the interior journey continues into the fen of the digestive system with 'those dark juices working' where, as for the Bog Queen, 'the seeps of winter/digested me' (Heaney 1980, 187).

The most sacred text for Irish Bogism is St Seamus' poem 'Bogland':

We have no prairies
To slice a big sun at evening--
Everywhere the eye concedes to
Encrouching horizon,

Is wooed into the cyclops' eye
Of a tarn. Our unfenced country
Is bog that keeps crusting
Between the sights of the sun.

They've taken the skeleton
Of the Great Irish Elk
Out of the peat, set it up
An astounding crate full of air.

Butter sunk under
More than a hundred years
Was recovered salty and white.
The ground itself is kind, black butter

Melting and opening underfoot,
Missing its last definition
By millions of years.
They'll never dig coal here,

Only the waterlogged trunks
Of great firs, soft as pulp.
Our pioneers keep striking
Inwards and downwards,

Every layer they strip
Seems camped on before.
The bogholes might be Atlantic seepage.
The wet centre is bottomless. (Heaney 1969, 47)

For Heaney, the bog and its processes of decay and digestion, death and life, are 'the vowel of the earth,' rather than the bowels of the earth. The consonants of the Earth are the hard, dry land, the harsh, sibilant sounds (like the souls of the sullen stuck in the slime of the Styx), the bones of the land, the exteriority of the Law of the Father, whereas the vowels of the Earth are the wetlands, the round, open sounds, the marrow, the breasts, the womb, the placenta, the interiority of the Body of the Mother, the grotesque lower bodily and earthly stratum, the place of life and death—living black waters.

7.2 H. P. Lovecraft: Weird Tale-Teller of the Bog

By valuing the bog as the quintessential Irish national and cultural (wet) landscape Heaney as the patron saint of bogs was following in the boggy and marshy footsteps of his fellow Irishman Lord Dunsany (1908, 31–45) in such modern fairy tales as 'The Kith of the Elf Folk' who live in an Irish marsh and in his first novel, *The Curse of the Wise Woman,* about an Irish bog. Heaney was also following in the boggy footsteps of H. P. Lovecraft in such weird stories as 'The Moon-Bog' also about an Irish bog. The Irish connection is also secured by the fact that 'The Moon-Bog' was written for a meeting of writers in Boston held fittingly on St Patrick's Day in 1921 and first published in 1926 (Lovecraft 2004, 409).

'The Moon-Bog' relates the story of a wealthy American of Irish background who moves back to Ireland to restore the ruined castle located by the ancient bog on the family estate. 'The peasants' are initially delighted with the American's restoration of the castle and they 'blessed him for bringing back the old days with his gold from over the sea' (Lovecraft 2004, 42). He was bringing the past back into the present which does not auger well as one never knows what might be brought back from the past into the present.

The American wanted not only to bring the past into the present, but also to destroy the past and to take the modern present back to the ancient past by draining the bog. Because of his foolhardiness, 'in time there came troubles, and the peasants ceased to bless him, and fled away instead as from a doom' (42). 'Doom' is a Lovecraft code word for apocalyptic destruction and the end of time. 'Troubles' is a term that refers in part to the troubles between the Catholics and Protestants in Northern Ireland.

The story can be read as a political allegory of the sectarian divide between the Catholics and Protestants in Northern Ireland, and between rural republican southern Ireland and industrial royalist Northern Ireland.

The American blames the bog as the cause of 'the peasants' fleeing and giving him trouble, but they left Kilderry because the American was 'to drain the great bog. For all his love of Ireland, America had not left him untouched, and he hated the beautiful wasted space where peat might be cut and land opened up' (43). Kilderry is a nice touch as it alludes to both the real city of Kildare and the famous Irish creamy ale Kilkenny. How Irish can you get?! The American has been corrupted by the modern drive to drain wetlands (as we saw critiqued earlier by Leopold) and wishes to do so in an act of agricultural sadism and geographical surgery out of misaquaterrism, the hatred of wetlands.

Unlike his friend the first-person narrator, the ugly American who hates the bog does not have an aesthetic appreciation for 'the beautiful wasted space' of the bog. The narrator relates earlier how he 'reached Kilderry in the summer sunset, as the gold of the sky lighted the green of the hills and groves and the blue of the bog, where on a far islet a strange olden ruin glistened spectrally. That sunset was very beautiful [...]' (42). The narrator paints a pastorally idyllic and picturesque depiction of the pleasing prospect in which the bog is not a blot on the landscape, or in the American's account books, but a vital, integral and beautiful part of it. The story can be read as an allegory of modern aesthetics that regards wetlands as ugly versus the traditional view of them as beautiful (as we saw critiqued earlier by Thoreau and Leopold).

The story can also be read as an allegory of modern capitalist resource extraction against the traditional love of bogs tied up with 'legends and superstitions' (43). When the American tells the first-person narrator about:

> the fears which had driven the people from Kilderry I laughed as loudly as my friend had laughed, for these fears were of the vaguest, wildest, and most absurd character. They had to do with some preposterous legend of the bog, and of a grim guardian spirit that dwelt in the strange olden ruin on the far islet I had seen in the sunset. There were tales of dancing lights in the dark of the moon, and of chill winds when the night was warm; of wraiths in white hovering over the waters, and of an imagined city of stone deep down below the swampy surface. But foremost among the weird fancies, and alone in its absolute unanimity, was that of the curse awaiting him who should

> dare to touch or drain the vast reddish morass. There were secrets, said the peasants, which must not be uncovered; secrets that had lain hidden [from past times long ago]. (43)

The ruined castle from the past could be restored to present glory, but 'the brooding bog' (44) in and of the past should remain there and not be brought into the present by being drained.

Yet this is precisely what the American was setting out to do. After the peasants flee, the American brings in 'labourers from the north' to drain the bog: 'now the work of drainage was ready to begin, and the labourers from the north were soon to strip the forbidden bog of its green moss and red heather, and kill the tiny shell-paved streamlets and quiet blue pools fringed with rushes' (44). The labourers from the north participate in the agricultural and voyeuristic sadism of stripping the bog of its clothing and in the American's aquaterracide (the killing of a wetland). The bog is personalized as a sentient being, if not feminized, the bride stripped bare by her haters (to adapt Marcel Duchamp).

The narrator is ambivalent about this act of environmental vandalism 'for although I disliked to see the moss and the heather and the little streams and lakes depart, I had a growing wish to discern the ancient secrets the deep-matted peat might hide' (45). He is driven by the same epistemophilia for enlightenment that drives his host. That night he awakes suddenly and sees out 'of a Gothic lattice window [...] a spectacle which no mortal, having seen it, could ever forget':

> The wide plain, the golden moonlight, the shadowy moving forms, and above all the shrill monotonous piping, produced an effect which almost paralysed me; yet I noted amidst my fear that half of these tireless, mechanical dancers were the labourers whom I had thought asleep, whilst the other half were strange airy beings in white, half indeterminate in nature, but suggesting pale wistful naiads from the haunted fountains of the bog. (45)

The labourers have been turned into uncanny automatons and the narrator is gripped by fascination and horror, twin affects of the uncanny.[1]

The narrator's ambivalence about draining the bog turns to dread: 'for some unknown reason I dreaded the thought of disturbing the ancient bog and its sunless secrets, and pictured terrible sights lying black under

[1] For Sigmund Freud. For further discussion of the uncanny drawing on his work, see Giblett (1996, 31–34), and on his and others, such as Friedrich Schelling, see Giblett (2019).

the unmeasured depth of age-old peat. That these secrets should be brought to light seemed injudicious [...]' (46). The enlightenment project of bringing the dark secrets of the past buried in the bog into the light of the modern present now seems ill-advised and highly dangerous. The American should have let sleeping bogs lie. What comes into the open in Lovecraft's story is the dangers of epistemophilia and enlightenment, and the evils of misaquaterrism and aquaterracide.

The following night the narrator has a similar vision, a recurring waking dream, or nightmare again seen through 'the Gothic window' (47) for the whole story is placed within a Gothic frame and seen through a Gothic screen or lens: 'as I watched in awe and terror I thought I saw dark saltant forms silhouetted grotesquely against the vision of marble and effulgence. The effect was titanic—altogether unthinkable—and I might have stared indefinitely had not the sound of the piping seemed to grow stronger at my left. Trembling with a terror oddly mixed with ecstasy [...]' (47). The narrator is experiencing the sublime of the lost city buried beneath the bog limned with the slime of the uncanny bog. Slime is the secret of the sublime as expressed in the portmanteau word s(ub)lime invented by Zoë Sofoulis and associated by her with the uncanny (cited by Giblett 1996, 27).

The narrator sees:

> Half gliding, half floating in the air, the white-clad bog-wraiths were slowly retreating toward the still waters and the island ruin in fantastic formations suggesting some ancient and solemn ceremonial dance. Their waving translucent arms, guided by the detestable piping of those unseen flutes, beckoned in uncanny rhythm to a throng of lurching labourers who followed dog-like with blind, brainless, floundering steps as if dragged by a clumsy but resistless daemon-will. (47–48)

The bog-wraiths seem to be gliding like flying creatures and floating like swimming creatures. They are half fowl and half fish, hybrids. They belong to neither the element of air, nor of water, but cross the liminal zone between them just as the uncanny automatons of the labourers cross the liminal zone between the human and non-human (dogs, daemons, machines).

The narrator:

> fled from that accursed castle along the bog's edge[.] I heard a new sound; common, yet unlike any I had heard before at Kilderry. The stagnant waters, lately quite devoid of animal life, now teemed with a horde of slimy enormous frogs which piped shrilly and incessantly in tones strangely out of keeping with their size. They glistened bloated and green in the moonbeams, and seemed to gaze up at the fount of light. (49)

Unlike most other Lovecraft tales, the smell of the stagnant waters is not described either for good or for ill. Typically, they would be described for ill as foetid odours, fishy or rank, but this story leaves the smell of the bog unremarked.

The narrator's final sight of the bog, however, is elaborated:

> I followed the gaze of one very fat and ugly frog, and saw the second of the things which drove my senses away. Stretching directly from the strange olden ruin on the far islet to the waning moon, my eyes seemed to trace a beam of faint quivering radiance having no reflection in the waters of the bog. And upward along that pallid path my fevered fancy pictured a thin shadow slowly writhing; a vague contorted shadow struggling as if drawn by unseen daemons. Crazed as I was, I saw in that awful shadow a monstrous resemblance—a nauseous, unbelievable caricature—a blasphemous effigy of him who had been Denys Barry. (49)

The bog with its monstrous frogs has its revenge against the monstrous American who wanted to drain it. Rather than the bog being monstrous, or being inhabited by a monster, the modern American is monstrous. The narrator's opening gambit for the story is to entice the reader with the enigma of 'somewhere, to what remote and fearsome region I know not Denys Barry has gone' (42). He has certainly gone into the bog, but presumably he has also gone into the regions far below the bog of the underworld never to return. He has received his just deserts for his misaquaterrism and aquaterracide in the triumph of good over evil and wisdom over folly, the moral of the story as in all good fairy tales.

7.3 Lord Dunsany and the Wise Woman of the Bog

Lovecraft was a great admirer of the writing of Lord Dunsany. Lovecraft's 'The Moon-Bog' precedes by five years the publication of Lord Dunsany's first novel, *The Curse of the Wise Woman,* also about an Irish bog. The first-person narrator describes the bog as 'the wild lands [...] where heather and mosses and rushes and soft black earth, and a million pools, stretched away out of my sight and knowledge' (Dunsany 2014, 11). On the following page the narrator reiterates and elaborates that 'the bog went on and on, over the horizon [where it seemed boundless (17)] and out of my knowledge' (12). The bog is out of his ken. It is a beautiful, but not necessarily, or always a safe place. 'The glimmering mosses [...] make the bog so beautiful' and safe from a distance, but the slime makes the bog dangerous so 'safe foothold and dangerous waters over unknown deeps of slime seem [...] to have been just about equal' (18). 'The quiet slime' is later claimed to be 'bottomless' (79). Heaney's poem 'Bogland' with its bottomless bog is following in Dunsany's footsteps. Heaney expressed admiration for Dunsany's novel in a piece written for *The Listener* and excerpted in the back-cover blurb of the Valancourt edition.

The bog stretches all the way into Tir-nan-Og, 'the land of the young' in 'its place outside geography [and] upon the outskirts of history' (73–74). Tir-nan-Og is beyond and before the bog. Both are spatially marginal on the edge of farming and mining land, and temporally liminal before farming and mining. 'Before' is used here in the double sense of being in the past and being in the future that will come to be present (in the present). Tir-nan-Og for Mark Valentine in his introduction to Dunsany's novel is 'the pagan Irish otherworld' (xi) of and beyond the bog. Dunsany contrasts 'the orderly woods inside what we called "the demesne"' with 'the other of things that began where man's cultivation ended' (12). 'The orderly woods' are later described as 'the tidy land lying park-like' (38) in front of the wilder woods at the far end of the demesne. Demesne (as we saw above in Chap. 9) is land on an estate reserved by the Lord of the Manor for his own personal use, such as for hunting, compared to land on an estate for the use of sub-tenants, such as for cultivation. Dunsany's narrator repeatedly contrasts 'the green and tidy country' of 'Earth subdued by man' with 'the rough lands' of the wild and watery 'waste' of the bog (60–61).

The orderly, park-like wood is the gentleman's park estate beloved of English landscape architects, as well as by explorers and colonists of Australia (see Giblett 2011, Chap. 4). English landscape architects loved to construct these parks out of the other-(wet)lands of bogs, fens and marshes. English explorers and settlers loved to find the gentleman's park estate in Australia as if God, or Nature, had created it, whereas Aboriginal people had made it. Where they did not find it, landscape architects loved to create parks and gardens out of the other-(wet)lands of bogs, fens and marshes.

Marlin, the son of the wise woman of the title, is not only the official 'bog watcher' for the estate appointed by the father of the first-person narrator (12), but also an indigenous Irishman for whom 'the bog was his country' (26). He is 'a kind of gate-keeper at the border of fairyland' (74) and 'clearly a citizen of Tir-nan-Og' (75) to which he eventually sets out via the bog on a journey from which there is no return: 'centuries hence, when we are all mouldered away, some turf-cutter will find Marlin there and will look on a face and a figure untouched by all those years, even as though the body had obeyed the dream [of Tir-nan-Og and going there] after all' (126). He has entered that 'strange eternity of the body' and become a 'bog-man' that Heaney celebrates. The narrator bemoans the fact that Tir-nan-Og and its bog-men are 'never mentioned in the Lives of the Saints' (149). They are mentioned now with Heaney as the patron saint of bogs, who, like all saints, follows in the boggy footsteps of those who have gone before, such as Dunsany, and is a devotee of the wise woman.

In Dunsany's novel, the Peat Development (Ireland) Syndicate mines peat from the bog and builds a dam, a factory, a mill race and houses for the workers. The foreman informs the narrator that 'they were going to compresses the peat, as he called what we call turf' (161). Peat is an inert and dead commodity, whereas turf is a growing and living entity; peat is Irish, turf is English; peat is modern, turf traditional. The modern peat-miners with their mechanized factory are contrasted with the traditional turf-cutters who manually 'brought the benignant influence of the bog to a hundred hearths' (116). Not only had the modern peat-miners: 'spoiled the magic that lay over all that land, deep as mists in autumn, but they were there for the purpose of cutting the bog away; not as turf-cutters take it, with imperceptible harvests, slowly, as years go by, a few yards in each generation, but working it out as miners work out a stratum of coal' (130). The turf-cutters are slow, sustainable harvesters on the surface of the bog, whereas the peat-miners are lower earthly stratum commodity-extractors in the depths of the bog.

The first-person narrator bemoans the fact that the Syndicate is mining 'the stratum of a lost era for which nobody cares. But who cares aught for the past? That pin-point of light called The Present, dancing through the endless night, is all that any man cares for' (118). The miners 'cut shafts' into 'the darker stratum' of 'the blackest layer' (161) that the peat had laid down in the carboniferous era, withdrawing on the bounty of the past and benefitting from it in the present in the chronological and geological time of the Anthropocene, and not in indigenous and mythical Irish now-time outside geography and on the outskirts of history in the symbiocene (as discussed in previous chapters and as coined by Glenn Albrecht, its patron saint).

Marlin's mother, 'the local seer, the prophetess Mother Marlin,' as Valentine calls her in his introduction to Dunsany's novel (ix), curses the miners, which sends the local Irish workers away, but not the English ones who are deaf to such heathen, pagan curses. 'The wise woman,' or 'practising witch' (22), warns that 'there's a power [...] that is hid in the heart of the bog that is against all their plans' (104). She is a personification of the bog and the bog is a manifestation of the wise woman. She is described earlier as 'somehow akin to those forces that ruled, or blew over, the bog' (15). She is of the same, extended family of the bog and the natural forces that rule over it. The curse of the wise woman is not so much the curses she utters as what she later calls 'the curse of the north wind' and 'the curse of the ancient powers of bog and hill' (147) that come back to haunt and monster the peat-mining operations.

The wise woman is akin to the forces that rule not only over the bog, but also under it described by the narrator as:

> the waters that Marlin called 'the sumach,' or some such word that I did not know how to spell, a mass of stored rains that grew heavier every year, till it flooded in under the roots of whatever growth gave a foothold, and floated the light surface of mosses and peat, till everything trembled round one as one walked: one called it the shaky bog, the most dangerous of all the kinds of bogs that one walks. (126)

Thoreau called such a shaky bog the swampy quaking zone. It is also the most dangerous of all the kind of bogs to mine.

The sumach is later described as 'the great store of the bog's water that kept all the mosses alive and their roots, and sustained and nurtured all that loved the bog, and made the steps of man unsure when he came, and made him come then as a stranger' (171). The sumach is a place apart from 'man,' an other place, the place of the other, an unhomely place that gives and takes life. When direct cursing does not send the English peat-miners away, the wise woman invokes the greater powers of the rain, the wind and the bog. The narrator describes how Marlin's mother 'was crooning to that horrible rain, and speaking to that fierce wind as though she were its equal' (171) and later how 'lovingly she spoke to the bog, bending down over the mosses, crooning to it and softly beseeching it, but what she said to it I did not know, for she was talking now that language that seemed older than Irish, which I had once heard her use before, and which certainly was no language that men speak now' (172).

She was speaking the language and singing the song of the boggy earth.

The wise woman also croons the will-o'-the-wisps, the bog lights: 'the great lights appeared, the lights I had often read of but never seen, the will-o'-the- wisps over the deeps of the bog; and strange as they looked out there on the desperate night, it was stranger to hear her crooning to them, welcoming them one by one, so far as I was able to make out from the tones of words that I could not hear' (173). Just as it would be prosaic and reductionist to call the sumach the aquifer beneath the bog, so it would be to call the will-o'-the-wisps ignited marsh gases as they are creatures of the bog capable of hearing and responding to the wise woman's crooning.

The wise woman croons the bog down into the heart of the bottomless, slimy depths of the strata of the sumach where the will-o'-the-wisps dwell. The strata of the past are crooned back into the present. The bog, 'that waste of moss and rushes and rushes,' goes 'on the march.' The bog is a body on the move. The beating, boggy heart of Ireland is living and the bodily bog of Ireland is moving. The bog 'arched itself into low hills.' The bog arches its back so that 'the whole bog was moving' (176). The English miners were deaf to the curses of the wise woman, but they cannot be deaf to 'the roar that the thing made [...] like a tide' (176), the curse of the bog itself that they escape just in the nick of time. The bog becomes a monstrous 'thing' whose 'roaring was louder than a tide; it was like a waterfall. The bog came grinding on, turning over and over [...] It covered the level land.' It also covered the houses and the factory, 'and still the bog roared on with the weight of all that mass of water' of the sumach behind it. 'And that was the end of the Peat Development (Ireland) Syndicate' (177). And that was also the end of Marlin's mother who joins her son in death in the bog to become a bog woman on their journey to Tir-nan-Og, the otherworld and after-world.

Rather than *The Curse of the Wise Woman*, the novel would have been more aptly titled *The Curse of the Bog* or *The Croon of the Wise Woman*. She is the crone who croons the song of the boggy body of the earth and the boggy heart of Ireland. Like Lovecraft, in whose story the bog itself stops the draining, in Dunsany's novel the bog revenges itself on the peat-miners through the song of the wise bog woman.

7.4 Kelly Barnhill: Singer of the Song of the Bog

Kelly Barnhill is another wise woman whose best-selling, *The Girl who Drank the Moon,* also sings a song of the bog:

> In the beginning, there was only Bog, and Bog, and Bog. There were no people. There were no fish. There were no birds or beasts or mountains or forest or sky.
> The Bog was everything, and everything was the Bog.
> The muck of the Bog ran from one edge of reality to the other. It curved and warbled through time. There were no words; there was no learning; there was no music or poetry or thought. There were just the sigh of the Bog and the quake of the Bog and the endless rustle of the reeds.
> But the Bog was lonely. It wanted eyes with which to see the world. It wanted a strong back with which to carry itself from place to place. It wanted legs to walk and hands to touch and a mouth that could sing.
> And so the Bog created a Body: a great Beast that walked out of the Bog on its own strong, boggy legs. The Beast was the Bog, and the Bog was the Beast. The Beast loved the Bog and the Bog loved the Beast, just as a person loves the image of himself in a quiet pond of water, and looks upon it with tenderness. The Beast's chest was full of warm and life-giving compassion. He felt the shine of love radiating outward. And the Beast wanted words to explain how he felt.
> And so there were words.
> And the Beast wanted those fit together just so, to explain his meaning. He opened his mouth and a poem came out.
> 'Round and yellow, yellow and round,' the Beast said, and the sun was born, hanging just overhead.
> 'Blue and white and black and grey and a burst of color at dawn,' the Beast said. And the sky was born.
> 'The creak of wood and the softness of moss and the rustle and whisper of green and green and green,' the Beast sang. And there were forests.
> Everything you see, everything you know, was called into being by the Bog. The Bog loves us and we love it. (Barnhill 2017, 83–84)

The Bog, like all wetlands, is a liminal zone between earth and water, between solid and liquid. This liminal zone is the most fertile zone on the planet where new life begins again and again. Long live bogs!

7.5 Rebirth in a Bog

Virgil had told Thorow him that if he really wanted to understand death and rebirth in a swamp he needed to go to Ireland and visit an Irish bog, such as the famous Black Bog. Virgil was still guiding them on the next step of their journey into an Irish bog. Virgil told Thorow that a long time ago people had been buried in the peat of Irish bogs where their remains were preserved intact. Their wrinkled bodies became dry and leathery, like dried up, polished shoes. Archaeology had brought them back to life in a way. They were buried in the bog and were reborn from the bog. They were a message from the past for the present and embodied hope for the future.

Sofia and Thorow decided to go on to Ireland as part of his study leave and visit Black Bog. A guide who took them around the bog told them about the wealthy American who had inherited the estate in which the bog was located. He came back to his ancestral home to find his family roots—or so he said. In fact, he wanted to uproot his family history and the natural history of the bog by draining it, mining the peat and selling it. When he tried to drain the bog, though, an old crone, or wise woman, who lived next to it crooned the bog so that it erupted and swamped the mansion of the American with him inside it. This story had come to be called 'the croon of the wise woman,' or 'the curse of the bog,' though it was more the American who had cursed the bog and it (or she, or both) was taking its revenge. He was alone in the house and was interred in it. Perhaps one day an archaeologist will find his leathery remains preserved in the bog.

As they were leaving the Visitors' Centre a gardener or ranger with 'Bog Keeper' monogrammed on his work cap and shirt rushed in and told the guide that an Australian worker was missing. She had not turned up for work that day and she had not slept at the backpackers' hostel last night. She might have got lost after she left the pub last night and disappeared into the bog never to be seen again.

Thorow could not help but overhear the conversation. He could not resist presenting himself as an Australian policeman, showing his Victorian Police ID and offering to assist. He could not help himself. He was a bit like a doctor on a plane when assistance is called for to attend to a sick passenger. The guide said that the missing gardener or ranger was an Australian backpacker so Thorow might be able to get some information from Australia or liaise with Australian police. In the meantime, he might be able to help until she was reported as missing to the local Garda Síochána and they took over the case. Thorow started asking the guide and other staff about the missing Australian backpacker. She was having a gap year between high school and university. She had been working on the estate for a couple of months doing casual maintenance work in the bog, like weeding, a never-ending job, and replanting native bog plants, a regenerative task. She was saving up enough money to continue her around the world journey before heading back to Australia at the end of her gap year.

Thorow got access to the hut where the gardeners or rangers stored their personal belongings while they were working. He found her passport in her backpack hanging on a nail. When he saw her surname and photo, he realized that she is the daughter of a very wealthy Australian global media tycoon with a finger in all the electronic and print pies in Australia and other countries. He also recognized her passport photo from when she had appeared in some family celebrity shots in the media when she was in high school. The bright star of a stellar family, or so the puff piece accompanying the story suggested, with a rosy future, destined for great things, and so on. Of course, the story and photo was published in a gossip magazine from daddy's stable and it was written no doubt by one of his own spin doctors so the result was hardly surprising. He must have wanted to give himself a human face by posing for the photo with his daughter and then wife, and not continue to appear to be the big, bad media robber baron he really was. Thorow remembered reading the story and seeing the accompanying photos in his doctor's waiting room. For some reason, the photo and story stuck.

He started to smell something fishy about her disappearance. Could she have been abducted and being held to ransom? If so, by whom? By a terrorist rogue cell of the IRA who were short of readies, or who wanted to make a statement about media monopolies, or both—kill two birds with one stone as it were? Or had she faked her own disappearance? If so, why? To get away from over-protective or over-ambitious parents? Or something else? Raped and murdered by an overeager Irishman or fellow backpacker who could not take 'no' for an answer after a night of too many Guinnesses at the local pub down the road and then dumped her body in the bog? This seemed like the most likely possibility.

Thorow decided he needed to go to the pub where she went last night and ask the publican and any locals who were there if they had seen anything untoward. Sofia and Thorow had been to the pub the previous evening. It was called unoriginally 'the Black Bog.' They had had a pleasant evening. Guinness, Kilkenny and pale ales from the local craft breweries flowed freely. The faces of the pub-goers were warm and glowing. The roast beef and potatoes were excellent washed down with a good claret. A log fire was burning brightly in the corner and the smoke from the fire and the various other substances being smoked created a warm and inviting place—a haven in a heartless world, a home away from home, a home where the heart is, or a hearth where the heart is anyway. It was a place to share stories and unburden oneself to a sympathetic listener. An Irish band played traditional folk songs and their own co-written radical songs taking the mickey out of Boris the Bozo, the uncanny Trumpster and other satirical targets. There was dancing and general merriment, a good, old traditional Irish pub.

The pub had just opened when Sofia and Thorow arrived. The publican welcomed them. He remembered them from the previous evening. The pub the morning after was a different scene and another story to the night before. It reeked of stale tobacco smoke that stained the walls and ceiling. It also stank of spilt beer that had soaked into the carpet and floorboards over many years. The inside of the pub this morning smelt like Great Grimpen Mire. Thorow and Sofia recalled the description from rereading it for his previous case. 'The Black Bog' sent an odour of decay and a heavy miasmatic odour into their faces. The tenacious grip of the pub was as if some malignant hand was tugging them down into its obscene depths, so grim and purposeful was the clutch in which it held pub-goers. Sofia was right that Conan Doyle's description of Great Grimpen Mire was lurid. She and Thorow managed to evade the pub's odiferous grip. The publican offered to serve them a drink. They declined. Too early for a drink. The pub was a dismal, damp simulacrum of the night before.

Thorow flashed his Victorian Police ID and said he was investigating the disappearance of a young, female Australian backpacker who worked on the Black Bog estate and who had just been reported missing. He happened to be visiting the estate at the time and offered to help until the Garda arrived. Yes, the publican knew the young woman. She came in regularly on a Friday evening after work. Last Friday, after a couple of pints, she had danced, not with anyone in particular, just a few different partners, male and female, and she left alone soon after. He didn't notice anyone following her. One guy had tried coming on to her a few weeks back, but she gave him such a knockback that he never came back. He was probably travelling through. He wasn't a local, the publican said. Never seen him before. She looked as though she could take care of herself. Tough little cookie, he said. Don't mess with her.

Thorow didn't think there was anything suspicious to follow up on here so he decided he and Sofia would go to where the young woman was last seen working in the Black Bog yesterday. He and Sofia headed off for the bright green bog under a cloudy sky. Thorow remembered the Irish taxi drivers in Dublin saying the sun never shines in Ireland. As a small island between the warming currents of the North Atlantic and the Irish Sea and with the prevailing winds, the location produces a mild climate, moist air, lush vegetation—and peat bogs.

Unlike Great Grimpen Mire and 'the Black Bog' the night after, Black Bog itself with its green-skimmed hollows, fresh-smelling quagmires, rank weeds and lush, slimy water-plants sent into their faces a pleasant, aromatic odour of decay, and of new life being reborn. No life is without smell and mess. The smell reminded Thorow of Erinmore tobacco from his younger days when he smoked a Peterson of Dublin 'Sherlock Holmes Baskerville'™ pipe (no kidding) and was pretending or aspiring to be the great fictional detective in real life. He had also bought a Deer Stalker from City Hatters in the basement below the famous clocks outside Flinders Street Station. He wore it home alone on the tram without getting any comments, a testament to Melbourne's tolerance of the weird and wonderful in fashion and clothing. Sofia heard about his escapade. She refused to let him wear it in public again and only allowed him to wear it at home in his study. He sometimes did when it was cold with the complete set of Penguin paperbacks of Sherlock Holmes novels and story collections arranged on his library shelves behind him. Might make for a nice cover pic for his memoirs one day, especially with him puffing away on the pipe in a smog of tobacco smoke and the Sherlock Holmes books in the background. Might help boost sales. Who knows? The reading public are fickle and like recycled heroes and stories. The picture would show him channelling the spirit of Sherlock Holmes, as it were, though how a living person might channel the spirit of a fictional character who had died when he plunged into the Reichenbach Falls clutched in an embrace with his arch-nemesis Moriarty, but had to be resurrected because of an outraged readership was a mystery beyond Thorow's, and probably also Holmes', consideration powers of deduction and solution.

Like Holmes and Watson sinking thigh-deep into Great Grimpen Mire, Sofia and Thorow occasionally sank boot-deep into the dark, quivering bog. The bog's soft embrace, when they sank into it, was as if some benign hand was tugging them down into its fertile depths, into its womb as it were, so warm and life-giving was the hold it had on them. Thorow recalled Thoreau's quaking zone.

Thorow found nothing suspicious where the young woman had been working yesterday so he and Sofia started scouting around the outskirts of the bog, down surrounding the lanes and byways to see if they could pick any clues or signs as to her whereabouts. It did not take them long to find an old woman sitting on a low, three-legged stool at an open fire outside her isolated stone cottage with a kettle singing away. The old woman was obviously in the middle of a long and intriguing story. Across the fire from her sitting cross-legged on the ground was the missing young woman Thorow recognized from her passport photo listening raptly to the old woman. Thorow was about to go up to them when Sofia put her hand gently on his forearm and said quietly, 'let me handle this.' Thorow nodded.

She approached both women who looked up and continued as before. She was obviously not threatening, whereas Thorow would have been and would have interrupted the story. Sofia pulled up a log and sat down with them and listened. Thorow hung in the background, making sure not to be obtrusive, but making sure that if the young woman did do a runner, he might be able to stop her, or at least speak to her. The three women were oblivious to him, engrossed in the story. After about 20 minutes or so, the old woman stopping speaking and poked the fire with a skewer. Some sparks shot up and all three women stared into the flames for a while in an obviously reflective mood.

Sofia then started talking to the young woman. Sofia gestured towards Thorow. The young woman turned around, started and seemed about to get up, but Sofia put her hand gently on her forearm and she subsided. Thorow could see Sofia talking quietly, but earnestly, to the young woman who listened carefully. Sofia stopped and the young woman talked to her.

Sofia eventually got up, came over to Thorow and said, 'she is sorry for any fuss or trouble she has caused by not telling anyone where she was last night after she did not go back to the backpackers. She stayed with the old woman last night because it started raining after she left the pub and it was too far to walk to where she is staying in the local backpackers hostel. She ended up soaked to the skin. She knocked on the door of the old woman's cottage who she had seen before working in her cottage garden and she invited her in. The old woman dried her off, sat her in front of the open fire, gave her a mug of hot chocolate with a dash of Irish whiskey and started telling her stories about the bog. The storytelling continued this morning as you saw.

'The old woman is the great grand-daughter of the old crone who crooned the bog into rising up and swamping the mansion when the American wanted to drain the bog and mine and sell the peat. There's more to the stories of the bog than the guide told us though. That was kind of a sanitised version for tourists. What the guide didn't tell us and what the old woman told the young woman is that the American was a good capitalist Christian who believed in his God-given mission of subduing the earth and getting money out of the land. This is what they call 'stewardship.' It's in the Bible somewhere. The bog represents the old, pagan, pre-Christian ways of Ireland. The old woman is the keeper of the old stories about the bog and the guardian of the bog itself, though she prefers to think of the bog as her mother who looks after her, rather than the other way around. This bog and all Irish bogs are the mother of Ireland. The bog is a sacred place. The American blasphemed the bog, the holy ground of Ireland, so he was cursed.' This was all starting to sound like an all too familiar story to Sofia and Thorow. They just nodded and smiled at each other.

Sofia continued, 'she told me that many Irish people who still believe in the old ways before Christianity came to Ireland come here on an annual pilgrimage to pay homage to the bog, to its patron saint and its other environmental apostles. Every year on the April 13 or 'Black Bog Day,' the birthday of the patron saint of bogs, they have a carnival on the estate with recitations of poetry about bogs by the patron saint and readings of stories by other environmental apostles who wrote about bogs. It's a fun time and the bog is a fun place. They have a marketplace where they exchange produce and crafts without using money. The carnival ends with a race in which competitors carry one's partner fireman style across their shoulders through the bog. For them, the carnival of Black Bog Day is the antidote to the black dog of depression. We have just missed it by a few days. I said we had been in Massachusetts on Patriot's Day. I explained what it was. I also said we had been Walden Pond where Thoreau, the patron saint of swamps, had lived. She was interested in learning more about Thoreau.

'She said we should have been here for the patron saint of bogs' day. Everybody had a great time. It is an Irish Patriot's day, like the American Patriot's Day. It is held ten days before St George's day on April 23 as a celebration of Irish independence from the England. These Irish get in first and celebrate the patron saint of bogs before the poms celebrate the patron saint of England. They also don't celebrate St Patrick's day on March 17 as he is the official patron saint of Ireland who converted the bog Irish to Roman Catholicism, whereas the patron saint of bogs converted them back to Irish Bogism. Celebrating the patron saint of bogs on Black Bog day a month after St Patrick's Day is the Irish having the final say and last laugh.' Thorow laughed too as the Irish loved to have the last laugh, especially against the English, just like Australians.

Sofia paused and Thorow asked, 'what about her—the young woman?' Sofia replied, 'oh, yes her. She wants to stay here in Ireland and live by the Black Bog and to learn more from the old woman. She also wants to learn more about the patron saints of bogs and swamps and the other environmental apostles. She wants to continue to work as a gardener and ranger on the estate looking after the bog and eventually to become a guide so she can tell the full story of the bog to visitors that the old woman told her.

'She does not want to go back to Australia and go to university and follow the path set out for her by her parents. She said that her parents are controlling and that they want her to go university and become a high-flying corporate lawyer. She said that is not what she wants to do with the rest of life. She did not know what she wanted to do until she came to Ireland and to the Black Bog and heard the old woman's stories. Her parents reluctantly agreed to her having a gap year so she could decide what she wants to do as a career, hoping that she would agree with them about becoming a lawyer. And she might have, she said, but not now. Her pathway is clear. She is going to stay here, learn more from the old woman and keep the stories alive. She wants to tell tourists the true stories of the bog, not the sanitized version.

'The old woman has no children so the stories could die out with her. The young woman hopes that she will become a kind of adopted daughter who will be entrusted with the stories and be able to pass them on. She feels that she has been reborn out of the bog. She is a new person. She says she has grown up and become a woman. I said, "a gap year can do that. You were too young to decide what you wanted to do with the rest of your life when you finished high school. You do now."

'I also said to her that she needs to tell her parents that she is alright and what she is going to do because they will be worried about her when the Garda contact them through the Australian Federal Police and tell them that she has been reported missing. She agreed to do that. Our work here is done,' Sofia concluded. Thorow nodded and they returned to the Visitors Centre and told the staff that they had found the young woman and she was safe and sound. She would be back soon to apologize for them thinking she was missing and for not letting them know she would be late for work that day.

References

Barnhill, Kelly. 2017. *The Girl Who Drank the Moon*. London: Piccadilly.

Dunsany, Lord. 1908. The Kith of the Elf-Folk. In *The Sword of Welleran and Other Stories*, 31–45. London: George Allen & Sons.

———. 2014. *The Curse of the Wise Woman*. Richmond, VA: Valancourt Books. First published 1933.

Giblett, Rod. 1996. *Postmodern Wetlands: Culture, History, Ecology*. Edinburgh: Edinburgh University Press.

———. 2011. *People and Places of Nature and Culture*. Bristol: Intellect Books.

———. 2019. *Environmental Humanities and the Uncanny: Ecoculture, Literature and Religion*. London: Routledge.

———. 2022. *Swamp Deaths: Collected Cold Cases and Other Marshy Mysteries*. London: Europe Books.

Heaney, Seamus. 1969. *Door into the Dark*. London: Faber and Faber.

———. 1979. *Field Work*. London: Faber and Faber.

———. 1980. *Poems 1965–1975: Death of a Naturalist, Door into the Dark, Wintering out, North*. New York: Farrar, Straus and Giroux.

Lovecraft, H.P. 2004. The Moon-Bog. In *The Dreams in the Witch House and Other Weird Stories*, ed. S.T. Joshi, 42–49. New York: Penguin. First published 1926.

CHAPTER 8

Marais Acadians in Canada

The first long-term European colonists in Canada were the French who settled briefly by the marshlands of the St Lawrence River at Cap Rouge from 1541 to 1543, and then settled the marshlands of Acadia on the Atlantic littoral from 1604 to 1755, with remnants returning later after the British deportation. They became known as the Acadians by virtue of settling 'Acadia,' roughly portions of the present-day provinces of New Brunswick, Nova Scotia and Prince Edward Island (for the origin of the name 'Acadia,' see Bible 1998, 18; Clark 1968, 71, n1; Herbin 1898, 14; Ross and Deveau 1992, 7–8). Using the small-scale technology of timber and sod construction that they brought with them from marshlands in France, the Acadians dyked and drained the rich and fertile soils in and around the Bay of Fundy in New Brunswick and Nova Scotia, most famously in the Grand Pré area. These soils were created by the tidal action of the Bay of Fundy as it has the highest tides in the world. These tides create not only fertile soil for humans, but also the ideal type of a temporally variable wetland for wader-birds with the land remaining wet after the ebb of the tide and never becoming completely dry as the tides ebb and flow on a six-hour cycle. The land is gradually inundated on the flow of the tide and exposed on the flow. The spatial and temporal intertidal zone is ideal habitat for wader-birds as they feed on crustaceans in the mud.

The intertidal zone also constitutes a permeable border between land and sea. A line marked on a map representing the edge of the land and the

R. Giblett, *Wetland Cultures*, Literatures, Cultures, and the Environment, https://doi.org/10.1007/978-3-031-57365-1_8

sea is a geographical and historical construction. Dyking created such a line and it is intended to maintain it to greater or lesser extent. The dyking technology of the Acadians did not try to create a bulwark against the sea (unlike that of the Dutch and British), but created a controlled, permeable barrier in which fresh water from rainfall was allowed to drain out from behind the dyke and salt water was prevented from intruding through the dyke, except occasionally (to which I return below). The line demarcating the edge of land and water is thus historically and geographically contingent on a wetland 'coast' on the north-western side of Nova Scotia with its marshes. This side of the province is entirely different from the south-eastern side with its rocky coast.

8.1 Acadian Grand Pré

The Grand Pré National Historic Site in Nova Scotia commemorates Henry Wadsworth Longfellow's fictional poetic character of Evangeline and her tragic love story in the eponymous long poem of that title, as well as the life and culture of the Acadians and their distinctive dykeland technology and farming techniques, including an impressive display of a recovered *aboiteaux*. Grand Pré is the setting for Longfellow's story. It is a distinctive dyked marshland as Longfellow points out early in his long poem, but this is not crucial to the story and it could have been set virtually anywhere, and in any period, though the style and genre in which it is told are quintessentially nineteenth-century, a century after the events on which it is based. Tragic love stories of separation and loss have been set in a variety of locales and times. The setting is not vital to the story and is not a character in its right with a part to play. In my story of these and other wetlands, they are characters in their own right, and have agency and a role to play. The Grand Pré Visitor Centre commemorates *Evangeline* and the Acadians, both their dyking and draining of marshlands, including Grand Pré, and their deportation by the British in 1755. The Acadians dyked and drained marshlands to manage grassland and thereby create farmland. Grand Pré was a marshland that was not amenable to European settlement until and unless it was dyked and drained.

Calling Grand Pré marshland, or even simply marsh, is to engage in a politics of naming. One hundred and twenty years ago, for Ganong (1903, 177, n.16), writing in a footnote about this area, 'the word *marsh* itself is rather a misnomer and is said locally to do the country some damage by giving an unfavorable impression of its character.' He does not go on to specify why the word marsh gives an unfavourable impression of the character of the country. Certainly, 'marsh' has a number of pejorative connotations such as dismal, melancholic, miasmatic and malarial. Some writers on marshes, most recently and notably John Stilgoe (1994, 47), have reconstructed the word 'marsh' as 'alongshore' in his eponymous book of this title in order to value its unique biology, physical features and aesthetic qualities and to ground what he also calls 'the marshland vocabulary' in the rich history of 'estuary English' to which he devotes a lovely small glossary essay.

Rather than marshland, Sheldon (1886, 24) used the term, or euphemism, 'grass land' when he described the Grand Pré area as 'a large area of what is probably the finest grass land on the American continent or in any other country.' Misperceiving and renaming marshland as grassland was a way of rendering it amenable to cattle-grazing and to dyking and draining, and so transforming it into farmland, another grassland, albeit an agriculturally managed and productive one. Seeing and naming marshland as grassland in the first place was a way of justifying its transformation into grassland by dyking and draining. Creating farmable grassland out of unfarmable marshland or grassland was merely a way of bringing to fruition the potential of the original marshland, or grassland. Grassland to grassland was the simple, tautological trajectory of development, except that the process belied the two different types of grasslands at the beginning and end of the process (marshland and farmland). Even naming the area 'Grand Pré,' literally 'Grand Prairie,' was a way of making it into grassland whereas it would have been more accurate to name it Grand Marais (Grand Marsh).

The dyking and periodic rejuvenating and flushing of the tidal marshlands became the quintessential landscape of the Acadians. These marshlands were transformed into what Bible (1998, 45, 48 and 76) calls 'rich meadow land.' 'Medow' and 'medowe' are also variously used in Henry Biggar's mock Tudor English translation of Jacques Cartier's journal of 1541 to describe the marshes of Cap Rouge as 'full of as faire and goodly grasse as ever I sawe in any Medowe in France' (Cartier 1993, 101). Bible goes on to relate that the marshlands of Acadia were 'reclaimed from the sea by the erection of dykes' and 'yielded most bountifully' as a benign Mother earth in the service of the dominant (agri)cultural paradigm and its patriarchal hierarchy. 'Reclaimed' is the stock-in-trade of triumphalist accounts of draining, as if the sea wrongfully claimed the land and the drainers rightfully reclaimed it back. Of course, the sea could always, and often did, reclaim the reclaimed land. Yet the sea, or more precisely its tidal action, created the bountiful land, or more precisely wetland, or marshland in the first place. According to Doucet (1999, 150), the Mi'kmaq calls 'the marshes the belly button of the universe.' They are a reminder of, and are a severed connection from, the nourishing maternal body of the earth, the 'pre-patriarchal' or matrifocal Great Mother earth in the alternative cultural paradigm, with the vestige of the umbilical cord that once connected them in symbiosis. Reclaiming the land from the sea created the patriarchal Mother earth, the dryland for agriculture in the service of the dominant cultural paradigm.[1]

As with a number of other Canadian wetlands (see Giblett 2014), Grand Pré attracted a posse of poets. Bliss Carman's first collection entitled *Low Tide on Grand Pré* was published in 1893. In the eponymous poem of the title first published in 1887, Carman (1893, 11) contrasts the 'barren reaches by the tide' (though they are in fact highly fertile) with 'the waving meadow lands' (that appear fertile and are fertile because of the tides). For Carman, 'a grievous stream, that to and fro/Athrough the fields of Acadie/Goes wandering' is also 'a drowsy inland meadow stream.' Carmen splits the stream into good and bad aspects. It is destructive and pastoral, malign and benign, bad and good in accordance with the moralistic binarism of the dominant cultural paradigm and its hierarchy that privileges the latter over the former. With the latter terms, Acadia is figured as Arcadia, as a pastoral idyll, and the Acadians and their work in making the meadow and building the *aboiteaux* that regulate the streams are rendered invisible. With the former terms, the tidal marshlands are rendered as 'grievous,' barren and infertile.

In the same vein and several years later, the poet and champion of the Acadians, and an Acadian himself, Frederic Herbin (1899, 16, lines 1–3), exclaims in 'The Dykes of Acadie':

> Oh, marshes green, the dykes of Acadie.
> I have been nursed upon thy ancient breast;
> I know the patience of thy lap's calm rest.

In Melanie Klein's terms, the dyked meadows are the good breast that nourishes; they are 'kind Nature' as Herbin (1899, 16, line 9) puts it in another poem. The undyked marshes are presumably the bad breast that does not nourish. The dyked meadows are also the womb out of which

[1] For further discussion of the dominant and alternative cultural paradigms, see Giblett (2011, Chap. 1, especially Fig. 1).

Herbin (1899, 16, line 6) is born, an autochthonous son 'new-born to life, by thee baptized and blest.' The dynamic and holistic ecosystem of land, water, air and weather has become split into moralized components. Presumably, the bad breast is too much water in the wrong place at the wrong time with breached dykes in storm surges and flooded farmland in heavy rains.

The tidal marshlands in the Bay of Fundy have played a vital role in its agricultural and cultural life and history. In a classic discussion of them, Andrew Hill Clark (1968, 28, 53–4) argues that they:

> largely supported Acadian people for more than a century and contributed so much to the economy of both Nova Scotia and New Brunswick for another century and a half [...] From the middle of the seventeenth century to the deportations of the mid-eighteenth, most of the Acadians depended primarily on the seasonally deep and fertile soils of the tidally flooded salt marshes.

They were able to do so by:

> the dyking of the marshes to exclude salt water, the cutting of drainage channels on their surfaces, and the construction of sluice gates (*aboiteaux*) [or what Clark later calls 'clappervalve gates' (238)], large and small, which allow the fresh water to drain out at low tide but exclude the salt water automatically when they close with the incoming tide. (Clark 1968, 30)

Despite this classic discussion, the pivotal role marshlands and dyking technology played in Acadian agriculture and culture have not always (and only recently) received the attention they are due by some Anglophone scholars and historians who, for instance, privileged the later Anglo-Celtic rocky southern coastal heritage over the earlier French-Acadian northern marshland coastal one, with the rocky Atlantic coast privileged as the quintessential Nova Scotian landscape over the grassy marshlands of the Bay of Fundy, except for a discussion of the fascination with the *Evangeline* mythology. *Aboiteaux* were also employed in the marshlands of the St Lawrence River, which have only recently received the attention they are due by some bilingual scholars and historical geographers, such as Hatvany (2003, especially 6). This divide in Canadian studies is symptomatic of the greater divides between English and French language speakers, between central and western Canada on the one hand, and eastern Canada on the other, and between rocky coast and grassy marshlands, in all of which the former is privileged over the latter.

For instance, in Ian McKay's (1994) *The Quest of the Folk,* an otherwise exemplary work of cultural materialist history, these divides are writ large. McKay discusses the Atlantic coast of Nova Scotia and the English and Scottish traditions rooted there, but largely ignores the Bay of Fundy, its marshlands and the Acadian-French tradition rooted *there,* with the exception of a couple of brief and tokenistic mentions of French settlement by the Acadians and their 'expulsion' in Anglocentric terms (McKay 1994, 26, 27, 229), the later captioning of pertinent photographs (269, 284) and a brief discussion of the *Evangeline* mythology (229). There seems to be for him a kind of Mason-Dixon line between the northern marshlands and southern rocky coast of Nova Scotia. In the twentieth century, in his view:

> the people of the fishing villages [on the 'South Shore' of the province from Halifax to Yarmouth (27)] came to be seen as bearers of Nova Scotia's cultural essence: they became archetypal [Anglo-Celtic?] Nova Scotians living not on a barren patch of the Atlantic shoreline but on a mythified 'Coast of Songs' [...] who led a 'simple life' in 'a rockbound landscape' and so 'draining their history of specificity.' (xv–xvi)

Yet equally in the twentieth century, the people of the farming villages on the north-western shores of the province from St Mary's Bay to Chignecto Bay came in similar ways and by similar means to be bearers of Nova Scotia's distinct French-Acadian cultural essence (albeit shared with their New Brunswickian and Prince Edward Islander fellow-Acadians) and so different from their Anglo-Celtic neighbours in both provinces: they became archetypal Acadians living not only on the marshlands of the Bay of Fundy, 'the Acadian heartland' for Ross and Deveau (1992, 21), but also in a mythified 'Land of Evangeline' where the Acadians led a 'simple life' in a grassy wetlandscape and so draining *their* history of specificity, just as McKay does here (but rectifies later and more recently with Bates in their *In the Province of History* [2010] to which I return below), and just as their dykes that drained the marshland, but rendering them into fertile, not barren, farmland.

What McKay (1994, 4) calls the mythification of the 'Coast of Songs' begins:

> on the margins: at the edge of the ocean [...] on the borderlands of land and water, the city and the countryside, the modern and the pre-modern [...] the dark, entrancing, romantic Otherness of the countryside [...] this liminal space.

Similarly, the mythification of the 'Land of Evangeline' begins in the moving margins of the intertidal zone of the Bay of Fundy, on the borderlands of land and water in the marshlands of Grand Pré, in the liminal space between the land and water of the wetland. Here, both the living black waters and rich black soils of the wetlands of the tidal marshes of the Minas Basin and others in the Bay of Fundy—with what Ross and Deveau (1992, 35) call 'their extraordinary fertility'—constituted the dark, entrancing, romantic otherness of the countryside. The Bay of Fundy 'and its numerous basins and inlets have the highest tidal ranges in the world—up to 15 meters in Minas Basin where the Acadians settled in the 1680s' (36). Grand Pré, literally 'great fields' (Doucet 1999, 144), is located in the Minas Basin and set in what le Blanc (2003, 13) describes as 'the most fertile agricultural area of Atlantic Canada' because of its marshlands and the native grasses edible by cattle.

The marshlands of Grand Pré for Ross and Deveau (1992, vii) are 'one of the most historically significant landscapes in Nova Scotia.' It would probably not be going too far to say that the dyked and drained marshlands of Nova Scotia and New Brunswick in general are the most historically significant landscapes for the Acadians. Or even to say that the dyked and drained marshlands of Nova Scotia and New Brunswick are the Acadian cultural and national landscape *par excellence*. They are, and were, their homeland, and their identity is deeply rooted in the lush grasses and wetlands of these marshes. They are significant, not least because they were one of the most important places for the dykeland farming that took place over a hundred-year period. Ross and Deveau (1992, 33) relate how 'the first dykes were probably constructed between 1640 and 1650.' They go on to relate that by, 'the end of the French regime in 1710, the Acadians had built dykes and *aboiteaux* in all the great marshland regions bordering on the Bay of Fundy' (36). Dykeland farming was for them 'to characterise the Acadian way of life in Nova Scotia until 1755' (21).

1755 was the year of the Grand Dérangement or Great Upheaval from the Acadian point of view, or the Great Deportation or Expulsion from the British point of view (Ross and Deveau 1992, 64). This event for Doucet (1999, 2) 'was an eighteenth-century precursor to ethnic cleansing.' For

Johnston and Kerr (2004, 6), it was 'ethnic cleansing,' albeit *avant la lettre,* before the term was invented. It was also an act of dispossession that permitted a new wave of Anglophone settlers to take over Acadian lands in Nova Scotia, New Brunswick and Prince Edward Island and to reap the benefits of Acadian dyking and draining.

8.2 Dykeland Farming

Dykeland farming, or what Ross and Deveau (1992, 33) later call 'lowland farming, which entailed the reclamation [*sic*] of salt marshes by means of co-operative dyking, became the single most distinctive characteristic of Acadian culture.' Or more precisely, dykeland farming was the single most distinctive characteristic of the Acadian culture of nature, specifically a second culture of nature that worked the land in labour-intensive agriculture using small-scale technologies (see Giblett 2011, Chap. 1). On this foundation, as Ross (2002, 9) puts it, 'the Acadians built a society based on the solidarity of mutual cooperation.' Acadian society, however, did not extend to, or embrace, mutual cooperation and solidarity with the non-human inhabitants of the salt marshes as dyking destroyed them and their habitat, nor did it respect the marshes themselves as dyking modified them highly.

The Acadians lived on the border not only between land and water, but also between the French and English empires. The Acadians for Griffiths (2005, 238), in her monumental study of them, 'lived the experience of a border people' located 'on the border between warring empires' and 'on a border between two great powers' in 'zones ruled by two different and competing empires' (Griffiths 2005, 184, 418, 438). They were borderland and marginal people living between empires and on the margin between wet and dry land. They were precursors for today's Canadian living on the border of empire to the south and in wetland country *par excellence.* Today's Canadians in their relation to empire are latter-day Acadians, and not just as near anagrams for each other. Just as empire cannot abide neutrality and wetlands that are betwixt and between land and water, so it cannot abide a people who live in this place and claim neutrality. The result was that the British deported the Acadians from their homeland with its distinctive dykeland technology and agriculture.

Dykeland farming was the single most distinctive characteristic of the Acadian culture of nature because it clearly and distinctively differentiated it from European settlements elsewhere in North America. For Ross and Deveau:

> no other group which colonised North America developed settlements based on the reclamation [*sic*] of salt marshes. No other group perfected the techniques of dyke-building. No other group preferred to settle on lowlands rather than clear the forests on higher ground. (1992, 35; see also le Blanc 2003, 27, 29)

In contrast to English colonization, which for Frederick Jackson Turner (1961, 45), in his famous discussion of the frontier, employed a farming frontier in North America (and cut down forests to do so), the French for Turner used a trading frontier (and did not cut down forests to do so). By contrast, with both of these frontiers, the Acadians used a distinctive dyking frontier with their *aboiteaux* (and cut down trees to construct them). Turner does not acknowledge the Acadian dykeland frontier, nor does Eccles, in the standard discussion of the Canadian frontier and of how it differed from the English one. Eccles (1969, 172) condescendingly describes and dismisses the Acadians as 'simple, peasant farmers,' whereas they were cooperative dykeland farmers using the simple, but effective, technology of the *aboiteau* and so were quite sophisticated. Their labour-intensive agriculture was potentially sustainable in the long term, though they never had the chance to prove this beyond one hundred years. No doubt their dyking also had a devastating and destructive impact on the marshland ecology and they did not live in mutually beneficial symbiosis with it and a lot of other, non-human creatures in their bioregion.

The Acadians differed from both the standard French trading and English farming frontier and their modes of colonization and cultivation. When it (or they) came to wetlands, the English employed a draining and filling frontier converting wet land to dry land suitable for dry land farming, whereas the Acadians used a dyking and reclaiming frontier transforming them into productive grasslands. The English also employed a clearing and tree-felling frontier when it came to forest or woodland as they often mistakenly thought these were fertile lands to create farmland whereas the Acadians preferred to drain and manage the productive and rich soils of wetlands and to avoid forests and their dangers. Greg Kennedy (2014) argues that the Acadians in part chose to live in coastal marshlands because they brought with them a fear of wolves from living in deep woods in western France. In fact, he has found proof of wolf attacks in the very parishes from which some of the colonists originated.

Rather than a frontier that implies a strict demarcation and a clear line between civilization and wilderness as Turner did, more recent work on European settlement in and of North America (such as McKay and Griffiths referred to above) and on Canadian culture (Jody Berland 2009) has employed the concept of borderland. This suggests a zone of interaction and interchange as well as a shifting and dynamic liminal space between, on the one hand, settlers and their settlements, and indigenous peoples and their lands on the other. The Acadians are certainly a case in point as they lived (on) the borderland between the French and English empires and lived on the border between land and water. Canadians today in general are a borderland people, too, as they live on the border of empire to the south and in a wetland country of borders of land and water. Canadian and Acadian are near anagrams for each other.

8.3 Aboiteaux

The Acadians carried out dyking in such a way that salt and fresh water were drained out on the ebb and salt water was allowed to occasionally 'enrich the marshes' (Ross and Deveau 1992, 36) on the flow, but prevented from constantly inundating the marshes. The central control mechanism of this small-scale technology and its agriculturally sustainable practices was the *aboiteau*, defined by Ross and Deveau as 'the wooden clapper valve that controlled the flow of salt water on the marshes' and that enabled them 'to exploit [and replenish] the natural fertility of the marshes' (35, 36). *Aboiteaux* for McKay and Bates (2010, 75) were 'marvels of early modern engineering' using sod, tree trunks as sluices and woven branches as the simple and sustainably deployed materials in their construction, and special sod spades as the tools for their construction.

Aboiteaux for Ross and Deveau (1992, 35) 'has no equivalent word in English or modern French,' though for Herbin (1898, 31) 'the word is doubtless of French origin' and for Ganong (1903, 177, n.16; see also Kennedy 2013) it is 'a word introduced by the Acadians from Saintonge, France, where it is still used in the form *aboteau*' [*sic*]. *Aboiteaux* contribute markedly to the construction and maintenance of Acadian cultural identity, distinct from both English and French culture, if not constituting Acadian 'exceptionalism.' The *aboiteaux* with their clapper valve were the distinctive feature of the Acadian wetlandscape of Nova Scotia.

Yet, as an aside here, Griffiths (2005) in her monumental study of the Acadians, neither mentions *aboiteaux* specifically by name as such, nor acknowledges their distinctive technological features, such as the clapper valve. Although she discusses dyking and draining (e.g. 67–69, 284–285, 310–311) and its important role in agricultural production where it was 'the key to the obvious success of Acadian agriculture' and in community building demanding 'a high level of cooperation' (132, 182; see also, e.g. 69, 311), she does not mention, describe, or use the word *aboiteaux*, a curious absence from her definitive 600-page study of the Acadians given the prominence they give to *aboiteaux* as a distinctive device in their cultural and agricultural history. Perhaps it is a case of not being able to see the trees (of the *aboiteaux*) for the wood (of Acadian dyking, draining and agriculture).

The process of mythification on both the rocky coast and grassy marshlands did not merely occur on the margins and in the borderlands of the city and the countryside, as if the two co-existed side by side in blissful harmony and mutual respect. Rather it involved what McKay (1994, 9) calls the 'aesthetic colonization of the country by the city.' The aestheticization of the country by the city was colonization by other means. As well as guns and germs, aesthetics and introduced plants and animals colonized native 'nature' and transformed land into landscape, both a visual phenomenon and a capitalist commodity (as Raymond Williams argued 50 years ago in *The Country and the City*; 1973, 120–125). This process had been going on for a long time before the twentieth century when there was both an aesthetic exo-colonization of the colonies by the colonial power and aesthetic endo-colonization of the country by the city both at 'home' and 'abroad.' Both sorts of colonization were also taking place in that most intractable and recalcitrant of places for aestheticization: the wetland.

Aesthetic endo-colonization of the country went hand-in-hand and sat cheek-by-jowl with tourism. McKay (1994, 33) notes the rise of the tourism economy and the tourism state in the 1920s and 1930s. This occurred equally (if not more so) in the 'Land of Evangeline' as on 'The Coast of Songs,' and the former had been occurring from the late nineteenth century as McKay (1994, 229) briefly notes in passing later and discusses at length recently with Bates (McKay and Bates 2010, 71–129). The development of the 'Land of Evangeline' as a tourist destination and site is explored in much greater detail by le Blanc (2003) in her appropriately entitled book *Postcards from Acadie,* beginning with the inclusion of a logo for the Dominion Atlantic Railway's 'Land of Evangeline Route' (le Blanc 2003, 60, Fig. 14) and culminating in 1926 with the formation of the 'Land of Evangeline Tourist Association' (McKay and Bates 2010, 71–98). Le Blanc's 2003 book seems to have provided the corrective stimulus for McKay, who takes up his earlier 1994 book and develops it in 2010 with Bates.

Despite the aestheticization and exploitation of both the rocky coast and grassy marshlands for economic and nationalist purposes, neither landscape was considered as nationalist. Canadian nationalism for McKay (1994, 36) 'fixated on central Canadian symbolic landscapes [such as lakes, forests and mountains] and historical myths created very little room for Maritimers' which should include both Acadian-French and Anglo-Celtic Maritimers, and quintessential Maritime landscapes, both rocky coasts and grassy marshlands.

8.4 Longfellow's *Evangeline*

The rise of tourism in the 'Land of Evangeline' in the late nineteenth century was due entirely to the publication in 1847 of Longfellow's long poem, *Evangeline*. Its opening line sets the scene for the first part of the poem 'in the Acadian land.' McKay describes *Evangeline* as 'an epic poem on the Acadian expulsion' (McKay 1994, 229). After the publication of the poem, McKay relates how 'Americans began to arrive in Nova Scotia to seek out the landscapes evoked by the poem' (1994, 229) in an early instance of literary tourism involving holding book in hand with landscape (and wetlandscape) before one seeking out the scenes described. The Currier and Ives 1864 print, 'Home of Evangeline-Acadian Land' (McKay and Bates 2010, 82–83), perfectly illustrated Longfellow's poem with its visual representation wherein, as McKay puts it, 'Acadia was clearly a pseudo-European Arcadia' (1994, 229; see also McKay and Bates 2010, 73) precisely located temporally in the period pre-'Expulsion' or pre-'Derangement.' The story of the fictitious character Evangeline for le Blanc (2003, 57) 'immortalized Grand Pré. For many readers, the poem also evoked images of an Eden on earth' albeit prior to the 'Expulsion.'

This vision of Arcadian, prelapsarian bliss contrasts with what McKay (1994, 284) calls the 'ersatz Acadian chapel erected at Grand Pré' to commemorate the Expulsion or Grand Dérangement (or Great Upheaval, or Great Dispersal) or Deportation, described by le Blanc (2003, 35) as 'the most traumatic and pivotal event [for many Acadians] in their collective history.' The British forcibly removed the Acadians for refusing to swear an oath of allegiance to the British Crown. The reasons for the deportation are complex, with political and military ones being the most compelling. Economic reasons were probably secondary and after the fact. Herbin claims that 'the great purpose of the deportation' was that the Acadians' 'land was offered to English settlers' (1898, 126). But, as Greg Kennedy (personal communication) points out, why burn everything if you just intend to steal it? The aftermath of the deportation certainly resulted in English and Scottish settlement of Acadia that took possession of their vacant lands and reaped the rewards of both Acadian dykeland farming and land ownership as the economic basis for agriculture.

The British transported the Acadians to the British American colonies. Some Acadians, as Herbin (1898, 129) argues, 'made their way to Louisiana and settled' there. This was not a British colony as Ross and Deveau (1992, 64) point out. It was Spanish. The Acadians were encouraged to settle there by the new Spanish authorities. The Acadians, as Greg Kennedy (personal communication) points out, were drawn by the opportunity to be autonomous and live in community, and not live under French control. They could thus continue to live their borderland life, albeit with another empire, this time the Spanish.

The chapel and statue of Evangeline erected at Grand Pré in 1922 and 1920 respectively were both commemorated in a Canadian 50-cent stamp of 1930 (le Blanc 2003, 127, Fig. 30) in perhaps the only instance of a fictional character being commemorated by a chapel, statue and postage stamp. Grand Pré thus became the capital and spiritual heartland of 'the Land of Evangeline' (McKay 1994, 269), though one should not forget 'the Land of Evangeline' in Louisiana represented in a painting of this title by Joseph Rusling Meeker (Miller 1989, Plate 2; Ellis 1998, 192) with its themes of what Ellis (1998, 192) calls 'loneliness and loss [...] in the mythic and melancholy Louisiana Bayou country,' stereotypical affects evoked by a romanticized wetland. For Ross and Deveau (1992, 63), Grand Pré was the place from which more Acadians were 'deported than from any other location,' not only because it was 'the most populated of all the Acadian settlements' but also because it was 'the most important Acadian agricultural and commercial centre.'

The views of Eden before and after the fall added poignancy to the story, and to the melancholy of the story and of the 'Land of Evangeline.' The pastoral perfection of, then expulsion from, the Nova Scotian 'Land of Evangeline' of Longfellow's historical romance segues into the tragedy of the Louisianan 'Land of Evangeline.' Both were wetlands, one the pastoral marshlands of Acadian Grand Pré and the other the swampy underworld of the Louisianan bayous that form a counterpoint to each other in the poem. Vileisis (1997, 101) notes that the result was that, 'even though Longfellow had never travelled in southern swamps himself [just as he had never travelled in northern marshes], his poem made a local Louisiana legend into the national rage,' not only for Americans, but also for Canadians. Equally, if not more so, his poem made a local Acadian legend into a Canadian national rage. Louisianans, and Americans more generally, claim the story originates from Louisiana, whereas Acadians claim the story originates from Acadia (le Blanc 2003, 53–55). *Evangeline* is a site of struggle between Americans and Canadians, between empire and border, for cultural and national ownership of the story. *Evangeline* is thus a site of struggle in cultural politics over what and how it shall mean.

Evangeline also manifests the taste for sentimental poetry prevalent in the nineteenth century. *Evangeline*, as New (1989, 38) remarks, 'was to become (for all its literary sentimentality and the plain fact of its American origin) a kind of talisman for the Acadians, who still resented the expulsion of 1755 and still preserved their separate culture.' One may dispute the claim for *Evangeline* being of exclusively American origin. Of course, its writer was American, but the tale has Acadian roots, certainly in its initial setting and main events, if not in its peoples. One may not be surprised that the Acadians 'resented the expulsion of 1755'—who would not? One may also dispute to what extent *Evangeline* is a marker of a separate Acadian culture. *Evangeline* was appropriated by mainstream Anglo-Canadian culture as a marker of Canadian national culture into, and by which, Acadian culture was subsumed. For instance, the first Canadian film of 1914 was *Evangeline* and the tourist documentary *The Land of Evangeline* was produced in 1936 (le Blanc 2003, 62, 96).

At the time of his death in 1882, according to le Blanc (2003, 52), Longfellow 'was considered the most popular poet writing in the English language and *Evangeline* is regarded by Zichy (1983, 241) as 'one of the most famous works in the English language [as it] appeared in some 270 editions and 130 translations between 1847 and 1947.' Longfellow is the only American poet to be honoured with a bust in the famous Poet's Corner in Westminster Abbey. A replica bust was unveiled at Grand Pré in 1955 (le Blanc 2003, 134–137, Fig. 31). Longfellow, le Blanc (2003, 98) concludes, was 'the best publicist the province could have ever had,' but he was a publicist for a sentimentalized construction of the province and its history that was the mere backdrop for the romantic story of thwarted young love that ended in tragedy—a real tearjerker.

Taking their cue from le Blanc and beginning where she leaves off, McKay and Bates (2010, 72) begin their recent discussion of *Evangeline* and Nova Scotia by arguing that 'the publication of Longfellow's masterwork ushered in a wholesale reorganization of an actual landscape in order to make it conform to a bestselling historical romance.' *Evangeline* was no mere publicity pamphlet for Acadia, nor a tour guide to it, nor a literary representation of a landscape and its history, but the template to which Acadia was cut. The statue of Evangeline at Grand Pré prompted one American tourist to wonder if she was buried around here somewhere (Ed Sulis, personal communication)! 'With only slight exaggeration,' McKay and Bates (2010, 77) suggest, 'one could acclaim the nineteenth-century Land of Evangeline as the very early beginnings of the twentieth-century theme park' entailing what they later call 'a thematization of landscape' (105). More precisely, this thematization was of a drylandscape and a parklandscape of Acadia, and of the wetlandscape of the Bayou 'with a maze of sluggish and devious waters' (Longfellow 2004, 64), and not of the wetlandscape of Grand Pré with its grid of flowing and controlled waters, nor of the Southern Bight-Minas Basin wetland of which it is a part.

Longfellow's Evangeline for le Blanc (2003, 51) 'symbolizes [...] the ideal Victorian woman.' Eve + angel = the angel Eve? In the Garden of Eden of Aca(r)dia, Evangeline is the prelapsarian Eve and not the later temptress of Eve who was the vector for Satan to tempt Adam into eating the fruit of the tree of the knowledge of good and evil. The British are the fallen angels who threw the Acadians out of the Garden of Eden. There is no Eve who is a temptress in this Garden of Eden, but there is the Eve of Evangeline who is expelled with her Adam Gabriel (also the name of a biblical archangel) from Eden. Evangel is also the 'Christian gospel' as McKay and Bates (2010, 86) point out. *Evangeline* is a secular gospel, though there is no good news in it.

The primary booster for the 'Land of Evangeline' in the later nineteenth and early twentieth centuries was John Frederick Herbin (le Blanc 2003, 115–118; Herbin 1898). In 1906 Bible (1998, 66) claimed that Herbin is 'the only living descendant of the Acadian exiles now residing in or near Grand-Pré.' Herbin had boldly made this claim himself on the title page of his 1898 book. Although he refers to Longfellow's 'beautiful poem' and asserts that it is 'a remarkably correct page of history' (1898,

7, 8), he is at pains later to point out the mistakes that he thinks Longfellow makes. Not least are the early lines (6–8) in which Longfellow describes how:

> Dikes, that the hands of the farmers had raised with labor incessant,
> Shut out the turbulent tides; but at stated seasons the flood-gates
> Opened, and welcomed the sea to wander at will o'er the meadows.
> (Longfellow 2004, 13)

Herbin (1898, 159) states emphatically that the 'gates are not opened to let in the salt water.' Yet Ross and Deveau (1992, 35), as we saw above, have suggested that salt water is occasionally allowed in to flush and replenish the meadows, whereas Ross (2002) later illustrates the *aboiteaux* to show them excluding all salt water, as does Bleakney (2004, Fig. 5.4, 52). Matthew Hatvany confirms the flushing thesis:

> In some of the Bay of Fundy sources, I have come across farmers who would periodically open the tidal gates for an entire season in order to allow the tides to deposit new sediment on the fields. This was a means of renewing the fertility of the fields and to raise the height of the fields which, after years of being dyked, would begin to compact, subside and become too low to permit good drainage. This technique was referred to as Warping, a European technique going back to the Middle Ages. (e-mail to the author, March 4, 2011)

John McLaughlin's 1995 Queens University PhD thesis on wetlands in Ontario discusses how the Acadians managed the sea meadows by dyking the lands (without using the term *aboiteaux*) in the tidal basins at the head of the Bay of Fundy (McLaughlin 1995, 105–106). He goes on to say, unremarkably, that the 'sluice gates in the dykes allow the fresh water creeks to pass through, but prevented tidal flooding.' He then says (remarkably in my view) that 'after about seven years of cropping, the gates were opened and the tides were allowed to pour in, replenishing fertility with nutrient-rich sediments; even though the land had to be left fallow for one to two years to allow the salt to leach out.' He then cites Clark's 1968 book on Acadia in support, without giving a specific page number.

McLaughlin's flushing thesis prompted a further exchange of e-mails with Matthew Hatvany, who cited Longfellow as above and then went on to argue that:

> he is clearly referring here to the purposeful opening of the floodgates to allow the sediment-laden waters of the bay onto the dyked fields in order to restore their fertility and raise the level of the dyked marsh. How often? My guess is that it wouldn't be less than every 20–30 years. Why? Because once a field is inundated by salt water it takes 2–3 years for the salt to leach out and the upland grass to recover. That is a long time for a farmer to lose the use of his/her field. (e-mail to author December 6, 2011)

Hatvany cited the famous 1903 article in support of this view in which Ganong states that 'this flooding, however, is by no means as extensively used as it should be, for many owners are unwilling, or cannot afford, to lose all return from their land for several years' (Ganong 1903, 178). Yet rather than referring to opening of the floodgates, Ganong is referring to breaking down of the dykes in order to flush the fields. Given the time and investment of labour in constructing them, this seems improbable.

More probable is the practice of propping open or removing the clapper valve to allow temporary and controlled incursion of salt water to replenish the soil and stimulate growth in the grasses. This seems to be the scenario Sheldon described in 1886 (referred to by Hatvany) which was called 'warping':

> [T]he dykes keep out the sea-water, which, however, as occasion may require, is let into the land, and escapes again through the *aboideaux* [*sic*]. When it is thus let in, it deposits a coating of finely granulated mud, which serves as a dressing of the best possible, manures, and operates for many years in this capacity, freshening up the land in an extraordinary manner. This system of warping is conducted with marked success in many places around the Bay of Fundy, on some of the land that was first dyked in by the French. (Sheldon 1886, 24–25)

Along similar lines, 'top-dressing' grass or lawn, is a well-known gardening practice to stimulate new growth of grass.

Hatvany's view is also confirmed by Harry Thurston (2004, 131–132; see also 170) in a recently documented account of his neighbour in northwest Nova Scotia who, as a salt marsh farmer of living memory, would:

> remove the clapper on the *aboiteaux* to allow the rides to flood his hay field, since the seeds of *Spartina patens* require soaking in salt water to germinate. This time-honored practice is called tiding and has a secondary benefit of renewing the fertility of the soil, as each tide brings a dressing of natural

fertilizer from the sea to the marsh. In this way the productivity of the salt marsh has been maintained for centuries, since the time of the Acadians, without need for manure or artificial fertilizers.

Longfellow thus got something right when it came to the Acadians and their dykes that have been amply acknowledged, but generally his otherwise saccharine account of Acadia has not been allowed to go unchallenged.

8.5 Acadia in Acadian Literature

Antonine Maillet's works for New (1989, 287) 'at once acknowledge and reject Longfellow's popular, familiar, but inauthentic "Evangeline" version of history.' Maillet's novel *La Sagouine* (first published in 1971), the confessional life and times of an older Acadian scrubwoman that began literary life as a one-woman play, is the de-romanticized counterpart to Longfellow's younger romanticized Acadian virginal woman (Maillet 2007). Along similar lines and unlike Longfellow who recounts the expulsion from Acadia, in Maillet's later novel, *Pélagie,* her first-person narrator tells the other end of the story, the return to Acadia as the subtitle of her novel indicates. Maillet was awarded the prestigious Prix Goncourt for this novel. She was the first Francophone writer not living in France and the first non-French citizen to be awarded this French equivalent of the Man Booker prize. In her acceptance speech she said, 'I have avenged my ancestors.' She did not say how. Perhaps she has avenged them against the British deportation and Longfellow's *Evangeline.*

The return to Acadie narrated in *Pélagie* begins by the Acadians avoiding floundering in Georgian swamps of 'Acadie-in-the-South' and ends up refounding Acadie in the Tintamarre (Tantramar) marshes of 'Acadie-in-the-North' (Maillet 2004, 11, 50). *Aboiteaux* mark the difference between the two places, and their wetlands, and their people. The novel begins and ends in wetlands and has its climax in a wetland too. Pélagie, the central character of *Pélagie,* wonders 'how had a place like Georgia drained its fields without aboiteaux [...] those wide dikes [in Acadie] that bordered the meadows and stole land from the sea [...] with their clappers opening and closing under the weight of the water' (23). The return is an exodus of Acadian exiles transported in 'the Cart of Life' pursued by 'the Wagon of Death' (11). The latter catches up with the former in the marshes of Salem where the Acadians struggle against both 'the sinking mud' and the Grim Reaper in the climax of 'the combat of the carts' (203, 211) set, like the novel's beginning and ending, in a wetland. This is a place of 'the earth moving' with 'nothing solid anywhere' and 'the earth itself caving in' (203, 205). In 'the sea of mud' (206) Captain Beausoleil builds a road of stones 'sunk in the mud' over 'the slippery mire' to try to extricate one of the carts from 'the ooze' (206, 211, 207, 208). At the same time he also has to extricate himself from the 'shifting mud that is trying to suck him down like a gaping mouth' (209).

Beausoleil's struggle is taking place in the quaking zone of 'the moving slime' of the marsh while in the meantime the centenarian Bélonie, who is rooted in the liminal and quaking zone of 'the clay of the marsh between earth and sea' of Acadie-in-the-North, repulses the Wagon of Death on her mission to claim an Acadian that day (213, 212). Pélagie eventually drives her off with the cry of 'Life' and so 'the Cart of Life [...] vanquished the Wagon of Death that day,' but not later when her cart becomes 'mired in the marshes of Tintamarre' (Tantramar) (215, 252, 252). Yet despite, or because of, her death, new life is born out of death and so with the 'coming home out of exile, Acadie was really born, like the first man, from a handful of mud' (252). Maillet rewrites the verse of the biblical book of Genesis that says God 'formed man from the dust of the ground.' In Maillet's rewriting of the biblical book of *Genesis* as a genesis from a wetland for a wetland people, the Acadians are born from the mud 'in the Tintamarre marshes [...] the cradle of the country' (254). They are not born in the marshlands of Grand Pré, nor anywhere near the 'formerly flourishing, lively town' (249) of Grand Pré, 'burned and deserted since that fatal day in September 1755. Out of superstition or fear of God, they hadn't dared settled anyone there' (249).

The marshes in the novel for Gann (1987, 31; all translations by Anusha Beechowa) are 'the prime place for the action in the story' as for him they form 'the dominant component of the landscape that Pélagie and her fellow travellers experience in their memory, in their present reality and in anticipation of their return.' The action shifts and time moves on, but the scene of action stays the same in the marshes and the past, present and future stay in the same place of the marshes. For Gann, 'as in a classical play, the exposure, the development of the action, the crisis and the outcome, everything happens in the same place, the marshes, no matter where they are located geographically' (31).

Gann notes the ironic connection between the wetland location of southern American swamps to which the British in 1755 deported some of the Acadians (such as Georgia and South Carolina), or in which they ended up by necessity or choice (such as Louisiana), and the Acadian marshes from which they were deported. He describes southern American swamps as 'the most unhealthy places in the American colonies, ironic reflections of the country from which they were torn away' (32). The former are more precisely 'the 'putrid swamps' of Georgia or South Carolina, the Coosawatchie, a kind of river that swamps the marshlands, making them uninhabitable' (32). These American swamps are unlike the Acadian marshes that were made inhabitable by the Acadians and their *aboiteaux*.

The antinomies of these wetlands are further adumbrated and nuanced by the American marshes in Salem. For Gann, 'the stages of their journey home are also outlined by this negative reflection of the maternal landscape, of their civilization: in North Carolina, in Virginia-Maryland and finally in Salem, the hostile marshes are present throughout the narrative.' More precisely, the hostile, unhealthy and death-dealing American marshes and swamps in the dominant cultural paradigm are posed against the nurturing, maternal and life-giving Acadian marshes in the alternative cultural paradigm. Gann goes on to comment that,

> It is mainly during the fight of the two carts, the clash between human heroism on one hand and cold death on the other that occurs in the very heart of the marshes of Salem, where the American marsh shows its significance. Far from nurturing and forming a new nation, this hostile landscape tries to swallow the carts and their carters. It is the last attempt of the Americans to absorb and assimilate these 'refuseniks' of the eighteenth century. (32)

Gann aligns the deported and exiled Acadians with the diasporic and wandering Jews, and the Soviet Union with the United States. 'Refusenik' is an individual who has been refused permission to emigrate. It was a term mainly used by the Soviet bureaucracy against Jewish people who wanted to migrate to Israel.

Gann goes on to comment that:

> Stubborn, the travellers refuse to concede that the marshes flooded by the spring rain are more dangerous than those in Georgia, the Carolinas and Virginia; the desire to move ahead towards their goal prevails over caution. Yet the winds and the sea shout their warnings, and become those of the screams of the witches of Salem or those of the shipwrecked in the imagination of the travellers. The landscape of sounds becomes a premonition. (32)

The screams of the Salem witches are a premonition of the soundscape of the Tantramar Marshes as the marshes are both life-giving and death-dealing. They claim Pélagie and Catoune when, as Gann puts it:

> a scream from Catoune 'that shook the dirty hay of the marshes of Tintamarre' announces his death. Since then, every windy night the voice of Catoune can be heard singing in the marshes of Tintamarre:
>
> there is song in these seasons. the
> Tintamarre. ghost birds over the
> centuries. voices in the tape of
> wind. caught to come back in their
> times. Tintamarre
> (Gann 1987, 33)

The Tantramar Marshes, like all wetlands, are both life-giving and death-dealing.

References

Berland, Jody. 2009. *North of Empire: Essays on the Cultural Technologies of Space.* Durham, NC: Duke University Press.

Bible, George. 1998. *An Historical Sketch of the Acadians, Their Deportation and Wanderings, Together with a Consideration of the Historical Basis for Longfellow's Poem 'Evangeline'.* Gretna: Pelican Publishing. First published 1906.

Bleakney, Sherman. 2004. *Sod, Soils and Spades: The Acadians at Grand Pré and their Dykeland Legacy.* Montreal & Kingston: McGill-Queen's University Press.

Carman, Bliss. 1893. *Low Tide on Grand Pré: A Book of Lyrics.* New York: Charles L. Webster.

Cartier, Jacques. 1993. *The Voyages of Jacques Cartier.* Toronto: University of Toronto Press.

Clark, Andrew Hill. 1968. *Acadia: The Geography of Early Nova Scotia.* Madison, WI: University of Wisconsin Press.

Doucet, Clive. 1999. *Notes from Exile: On Being Acadian.* Toronto: McClelland & Stewart.

Eccles, William. 1969. *The Canadian Frontier 1534–1760.* New York: Holt, Rinehart and Winston.

Ellis, Amy. 1998. Joseph Rusling Meeker 1827–1889. In *New Worlds from Old: 19th Century Australian and American Landscapes,* ed. Elizabeth Johns, Andrew Sayers, and Elizabeth Kornhauser, 192. Canberra: National Gallery of Australia/Hartford: Wadsworth Atheneum.

Gann, Andrew G. 1987. Pelagie-La-Charette: Paysage Interieur, Paysage Exterieur. In *The Red Jeep and Other Landscapes: A Collection in Honor of Douglas Lochhead*, ed. Peter Thomas, 28–33. Sackville: Centre for Canadian Studies, Mount Allison University/Fredericton: Goose Lane Editions.

Ganong, William. 1903. The Vegetation of the Bay of Fundy Salt and Diked Marshes: An Ecological Study. *Botanical Gazette* 36: 161–186.

Giblett, Rod. 2011. *People and Places of Nature and Culture*. Bristol: Intellect Books.

———. 2014. *Canadian Wetlands: Places and People*. Bristol: Intellect Books.

Griffiths, Naomi. 2005. *From Migrant to Acadian: A North American Border People 1604–1755*. Montreal & Kingston: McGill-Queen's University Press.

Hatvany, Matthew. 2003. *Marshlands: Four Centuries of Change on the Shores of the St. Lawrence*. Sainte-Foy: Les Presses de l'Université Laval, Les Èditions de L'IQRC.

Herbin, John Frederic. 1898. *The History of Grand-Pré: The Home of Longfellow's 'Evangeline'*. 7th ed. Wolfville: Herbin Jewellers.

———. 1899. The Marshlands. In *The Trail of the Tides*, 2nd ed. Toronto: William Briggs.

Johnston, John, and Wayne Kerr. 2004. *Grand Pré: Heart of Acadie*. Halifax: Nimbus.

Kennedy, Gregory. 2013. Marshland Colonization in Acadia and Poitou During the Seventeenth Century. *Acadiensis* 42 (1): 37–66.

———. 2014. *Something of a Peasant Paradise? Comparing Rural Societies in Acadie and the Loudunais, 1604–1755*. Montreal & Kingston: McGill-Queen's University Press.

Le Blanc, Barbara. 2003. *Postcards from Acadie: Grand Pré, Evangeline and the Acadian Identit*. Kentville: Gaspereau Press.

Longfellow, Henry Wadsworth. 2004. *Evangeline: A Tale of Acadia*. Fredericton: Goose Lane Editions. First published 1895.

Maillet, Antonine. 2004. *Pélagie: The Return to Acadia*. Trans. P. Stratford. Fredericton: Goose Lane Editions.

———. 2007. *La Sagouine*. Trans. W. Grady. Fredericton: Goose Lane Editions.

McKay, Ian. 1994. *The Quest of the Folk: Antimodernism and Cultural Selection in Twentieth-Century Nova Scotia*. Montreal & Kingston: McGill-Queen's University Press.

McKay, Ian, and Robin Bates. 2010. *In the Province of History: The Making of The Public Past in Twentieth-Century Nova Scotia*. Montreal & Kingston: McGill-Queen's University Press.

McLaughlin, John. 1995. Progress, Politics and the Role of Conservation: Wetland Drainage in Ontario. PhD thesis, Kingston, Ontario: Queen's University.

Miller, David. 1989. *Dark Eden: The Swamp in Nineteenth-Century American Culture*. Cambridge: Cambridge University Press.

New, William. 1989. *A History of Canadian Literature*. Basingstoke: Macmillan.
Ross, Sally. 2002. *Dykes and Aboiteaux: The Acadians Turned Salt Marshes into Fertile Meadows*. Grand-Pré: Société Promotion Grand-Pré.
Ross, Sally, and Alphonse Deveau. 1992. *The Acadians of Nova Scotia: Past and Present*. Halifax: Nimbus.
Sheldon, John. 1886. *To Canada, and Through It, with the British Association*. Ottawa: Department of Agriculture.
Stilgoe, John. 1994. *Alongshore*. New Haven, CT: Yale University Press.
Thurston, Harry. 2004. *A Place Between the Tides: A Naturalist's Reflections on the Salt Marsh*. Vancouver: Douglas & McIntyre.
Turner, Frederick Jackson. 1961. The Significance of the Frontier in American History. In *Frontier and Section: Selected Essays of Frederick Jackson Turner*, 37–62. Englewood Cliffs, NJ: Prentice-Hall. First published 1893.
Vileisis, Ann. 1997. *Discovering the Unknown Landscape: A History of America's Wetlands*. Washington, DC: Island Press.
Williams, Raymond. 1973. *The Country and the City*. London: Chatto & Windus.
Zichy, Francis. 1983. *Evangeline*: A Tale of Acadie. In *The Oxford Companion to Canadian Literature*, ed. William Toye, 241. Toronto: Oxford University Press.

CHAPTER 9

A Marsh Writer in Canada and the Future of Wetland Cultures

The native Nova Scotian and self-confessed nature writer Harry Thurston loves marshes and writes lovingly about them. Thurston grew up near one marsh in south-west Nova Scotia and now lives near another one in north-west Nova Scotia. He writes appreciatively and extensively about both of them, often linking them together as two important locales and the book-ends of his life as a boy and a man. His first encounter with a marsh was when he was a boy and he went to a marsh close to the family farm near the Chebogue River in south-west Nova Scotia. In 'Tidal River,' the first section of his 'Arcadia: The Marsh Suite' of poems, Thurston (2000, 62; see also 2004, 3; 2009, 48) translates '*Che-bo-gue*' as '*Great-still-water.*' This river, Thurston (2004, 3) says, is 'twice daily swelled with tide, nourishing the great salt marshes bordering it. These were and remain today the largest intact salt marshes in Atlantic Canada.' They have not been dyked, drained, backfilled or 'infilled' (to use Thurston's terms), unlike the vast majority in Atlantic Canada (as we have seen in the previous chapter and as Thurston relates elsewhere too as we will see below).

Since 1990, Thurston has lived overlooking a marsh near the family home next to the Tidnish River in north-west Nova Scotia close to the provincial border with New Brunswick. Thurston's two marshes were connected for him aurally rather than visually, the later one with the earlier one, over the span of 30 years by the call of the willett, a large migratory shorebird. On the first morning in his new home by the Tidnish marsh,

R. Giblett, *Wetland Cultures*, Literatures, Cultures, and the Environment, https://doi.org/10.1007/978-3-031-57365-1_9

Thurston (1996, 1) hears 'a familiar, and deeply affecting, sound' that to him 'could be no clearer signal that I was indeed home' because the sound transported him instantaneously and imaginatively to 'the marsh environment of my childhood.'

9.1 The Salt Marsh is 'My Home'

Thurston has a strong sense of home, not just in his house, but also in the home-habitat of his local place shared with other living beings. Indeed, Thurston (2004, 6 and 7) calls the salt marsh 'my home' and he realizes when he returns to it, albeit located elsewhere in Nova Scotia, that he is returning to 'my original habitat.' Thurston (2004, 214) later describes how, 'when we moved to our present home, with its view of the Old Marsh, I was returning to the landscape of my childhood,' even though he was not returning to the marsh of his childhood. At the very end of *Between the tides*, Thurston (2004, 231) reiterates that 'in midlife, after a circuitous journey, I had found my way back to the environment of my childhood,' even though again he was not returning to the specific marsh of his childhood. Thurston's attachment to salt marshes is transportable; his sense of place changes from one place to another, albeit over longish periods; his environment, habitat and landscape are the salt marsh wherever it is. Wherever the salt marsh is, there is his home. Home is where the salt marsh is. The salt marsh is his heart.

Against the possible charge that he is nostalgic for the lost innocence of his childhood and its play places, Thurston (2004, 214) counters by insisting that 'my attachment to the salt marsh is more than nostalgic or cultural. It is visceral and instinctive [...] this is my habitat.' Thurston (2004, 228) relates later how he did return to his childhood farm and marsh and how he found the farm and its surrounds changed, but the marsh was 'as verdant and vital as it had always seemed. Far off, I heard the cry of a willet.' This cry that later transported him from his new home at the Old Marsh at Tidnish back to his old home at this marsh, now reaffirms his attachment and commitment to the marsh as his and the willet's home. The place of his childhood had changed as the brook was dead as it had no fish, 'but the salt marsh, was for now, still a living place, still home to the birds of my childhood, still a place of wonder.' Thurston's sense of home includes non-human species; home is shared with others— his sense of place, a sense of wonder.

Thurston is a bioregionalist. Although Thurston does not use the term 'bioregion,' for him his current home-habitat is not just his local place but the watershed, including the larger Tantramar Marshes of which the smaller Tidnish marsh is a part, the two rivers of these names, one flowing south to Chignecto Bay and the other north to Baie Verte, and the tidal action of the Bay of Fundy. Like Henry David Thoreau, who kept a nature journal and traced the seasons in a phenology (writing the seasons) of the woods, ponds and swamps, Thurston (2004, 7) kept 'a marsh journal' and compiled 'a phenology of the salt marsh' that traced its life through the seasons. Thurston quotes from Thoreau's work periodically in *A Place between the Tides*, principally from *Walden*, though this is one of Thoreau's least wetland-oriented writings. Thoreau's more wetland-oriented writings are in his essay 'Walking' and in his *Journal*. Like Thoreau, Thurston (2004, 61) regards himself as a naturalist as he, and the subtitle of *A Place between the Tides – A Naturalist's Reflections on the Salt Marsh*, indicate. But he is not just any old naturalist or weekend hobbyist or bird twitcher, but the naturalist whom he qualifies as one who 'puts down roots as I have done,' just as Thoreau did next to Walden Pond and in Concord County.

Thurston (2004, 159) later calls himself, or describes how he is regarded locally as, 'the community naturalist and environmentalist.' For Thurston (2004, 224), being the community naturalist does not refer to just the human community but also to 'a larger community' of 'my fellow "biotic citizens"'—to borrow the beautiful phrase coined by the great twentieth-century conservationist, Aldo Leopold. Rather than addressing 'my fellow Americans' as the US President does in his annual 'State of the Union' address, Leopold as the inaugural President of the US Wilderness Society addresses 'my fellow biotic citizens' in an ongoing 'State of the Earth' address to the 'parliament of all beings.' In *A Sand County Almanac*, he developed the concept of the larger community. In the 'Foreword,' Leopold (1949, viii) suggests that 'when we see land as a community to which we belong, we may begin to use it with love and respect.' In other words, when we begin to use the land as community with love and respect, we develop and practise a land ethic. In his essay on 'The Land Ethic' he discusses 'the community concept.' Leopold's (1949, 204):

> land ethic simply enlarges the boundaries of the community to include soils, waters, plants, and animals, or collectively: the land [...] a land ethic changes the role of *Homo sapiens* from conqueror of the land community to plain member and citizen of it.

Leopold's land ethic is a biocentric point of view and practice of the conservationist who sees humans as part and parcel of the land and the cosmos.

Thurston's *Between the Tides* is a kind of Canadian 'Salt Marsh Almanac' to place alongside its American counterpart Leopold's *A Sand County Almanac*. In an interview with Lisa Szabo-Jones (2011, 85), Thurston said, 'what I'm dealing with is the almanac which is the cycle of the seasons.' The first part of Leopold's book is organized by the twelve months of the year, as is the whole of Thurston's. The second part of Leopold's book is a kind of travelogue, including his account of his visit to the Clandeboye Marsh in Manitoba. Similarly, Thurston occasionally includes autobiographical anecdotes about his travels elsewhere in Canada and what he learnt from them. The last part of Leopold's book contains his ground-breaking and now classic essays on conservation aesthetics, land ethics and wilderness. *Between the Tides* has no equivalent to this part, but it does reflect, as the subtitle indicates (and as we will see), on the salt marsh as it goes along. This reflection leads to a discussion of some general ecological concepts and conservation issues. In this regard, it is more in the tradition of Thoreau's *Walden* than Leopold's *Sand County Almanac,* though, unlike Thoreau, Thurston does not discuss the economics of living by Tidnish marsh other than to divulge that he is a househusband and that his wife Cathy is the main breadwinner. The economics of living in a place is an important and vital facet of any consideration of living in a bioregional home-habitat. The ecological and the economic are not just linked by the common prefix of 'eco,' or household, but also by the association of goods and services provided in and by a bioregional home-habitat that sustain life and livelihoods, such as that of writers like Thurston who presumably derives his livelihood in part from writing about it.

Being a community naturalist who has puts his roots down has its rewards and responsibilities for Thurston as 'a de facto bird rescuer.' Thurston (2004, 85–86) appreciates 'the reward for staying in one place and keeping watch. Travel yields diversity; residence, intimacy. Staying in one place, looking and looking while the seasons rotate around you, reveals the patterns inherent in the familiar.' Thurston (2004, 146) even later likens these patterns that accompany 'the turning of the seasons, especially the appearance and disappearance of certain species,' to 'the finely tooled workings—the escapement—of a pocket watch.' Yet there is nothing as predictable and mechanical as clockwork in seasonal patterns as this Cartesian view of nature as machine would suggest. Indeed, this analogy contradicts Thurston's (2004, 61) earlier argument that 'there are two distinct kind of pleasures that derive from being in one place over time and tracking the rounds of the seasons. The first relates to the expected [...] The second relates to the unexpected.' The unexpected is not predictable like clockwork.

Thurston derives bodily, including sensory, pleasure from being a naturalist in performing the principles and practices of Leopold's land ethic—Thurston (2004, 84) relates how he adopted 'a proper conservation ethic'—and appreciation for the sight and sounds (and smells and touch?) of land, living beings and natural processes that Leopold called a conservation aesthetic. For Leopold, land ethics and conservation aesthetics went hand in hand, and one was not viable without the other. Without conservation aesthetics, a land ethic is a painful set of dicta and rules leading to an ascetic religion of denial and loss; with conservation aesthetics, a conservation ethic is a pleasurable practice of being in place and enjoying an appreciation for the sight, sounds, smell, taste and touch of land, living beings and natural processes in the cycle of the seasons.

Thurston is not what Leopold also called an alphabetic conservationist, nor is he a twitcher. Thurston (2004, 76–77) states categorically:

> I am not a 'twitcher,' that dedicated breed of birder who, for example, frequents sewage lagoons in search of a rare bird. And I have never kept a list, except for the informal one that has emerged from my marsh journal. But I must confess to a certain satisfaction in knowing where and at what season to find certain species.

'The circle of the seasons' is reassuring for Thurston (2004, 77) as the circle is 'revolving as it should' and 'the pattern underlying the cycle of life is intact.' Thoreau had a similar appreciation for the round of the seasons, for the changes they brought and for the reassurance they gave (see Giblett 2013, 149–163).

Thurston began to appreciate the cycle of the seasons and the rhythms of nature when he was a boy. Thurston (1990, 8) begins his 'Prologue' to his book *Tidal Life: A Natural History of the Bay of Fundy* by relating how 'the tides of Fundy first entered my life along the umbilical loops of the Chebogue, a tidal river at the southwestern tip of Nova Scotia.' The umbilical loops of the river connect the watershed to the sea through the placenta of the marsh. Later, the tidal Tidnish River is for Thurston (2004, 215) 'the salt marsh's umbilicus to the mothering sea.' Yet the sea is not as fertile as the estuary, nor is the estuary as fertile as the marsh. The marsh is mothering and the marsh is the mother who Thurston (2004, 215) anthropomorphizes with 'the dark eyes of the marsh pools.' An estuary for Thurston (2004, 174), 'where salt and fresh waters mix, is ten times as productive as the open ocean, and the salt marsh is ten times as productive as the estuary.' The sea is not mothering at all; the marsh is mothering to all. Earlier he had experienced the 'feeling of a boy in the midst of the marsh, the green world enfolding me' (Thurston 2004, 114). The marsh is the great mother that embraces and nurtures the child, the father of the marsh man. The marsh is an enfolding womb with a nourishing placenta. Thurston (1990, 8) relates how 'the tidal creek and salt marsh that bordered my father's farm also introduced me to the rich and varied natural life of Fundy.' The marsh's placenta give birth to new life in the wide-open world of the air and water of the Bay. The marsh is mother.

In 'Arcadia: The Marsh Suite,' Thurston (2000, 62) writes how he returned to the Chebogue, 'the place/of childhood' where he is 'an eye still rejoicing/at the flat, green world/leavened with salt.' The flat green world of the great still water rejoices the eyes and salt leavens this flat world so it rises to be the staff of life. He does not describe how salt also addresses the sense of taste, closely related to the sense of smell, by enhancing flavour and adding piquancy. Thurston generally privileges the sense of sight and the organs of the eyes over the other senses and the other sense organs. Indeed, Thurston (2004, 12) calls himself 'the marsh watcher,' whereas Thoreau was a swamp wader, though Thurston (2004, 125) also refers to 'us as marsh dwellers.'

That company also includes Thoreau. Like Thoreau, he also enjoys 'the sensuality of being in the field—of seeing, feeling, or even tasting a species where it lives' (Thurston 2004, 134). Perhaps, moreover, even smelling where it lives. Later, at the end of *Between the Tides*, he insists that 'my enduring attachment to the animals and plants is far more sensual than it is cerebral' (Thurston 2004, 225), just as it was for Thoreau, and just as it was a multi-sensual attachment to home place for Thoreau.

Thurston needed to unlearn his formal, cerebral education and to learn to use and trust his own senses in his informal, sensual education. Similarly, Leopold (1949, 18 and 158) posed the question 'is education a process of trading awareness for things of lesser worth?' Later he answered his own question. He feared that education is 'learning to see one thing by going blind to another,' including 'the quality of marshes.' Thurston (1990, 11; 2004, 110) studied Biology at Acadia University in Wolfville, Nova Scotia, in 'the heart of the Land of Evangeline' as he says, neither of which seemed to have contributed much to his seeing marshes despite their history and location as he does not go into any further detail about the history and ecology of the marshlands of Grand Pré. Perhaps he feels they have been done to death by other writers. The literature on Grand Pré is certainly extensive (as we saw in the previous chapter of the present volume).

Later Thurston (2004, 8) says his 'formal education [was] as a biologist' and his 'informal education [was] as an environmentalist.' The former certainly privileges the sense of sight as the perceiver and arbiter of truth to see, and to know through and by seeing. The latter for Thurston 'sharpened' his sight and this gave him 'a more wary eye' to view the 'pageant of life being played out between the land and the sea.' Yet this pageant is not only visible to the eye, but is invisible to it too. As Thurston (2004, 107) says later, 'so much of life's drama goes unseen' (and unheard, and so unappreciated on both counts, by both aesthetic senses; for Hegel seeing and hearing were the only legitimate aesthetic senses). This pageant also addresses the other senses in the smell of its odours, the taste of its foods, the touch of its textures and the sound of its calls and cries, such as the willet's cry. For Thurston (2004, 14), unlike other mammals for whom 'smell is the most important sense, we rely most on sight.' If 'we' refers to modern, urban dwelling humans who use ocular communication and other visual technologies and have done so beginning at about the time of the Renaissance with the development of precise lenses and the invention of the telescope and later microscope, the sextant and theodolite, then 'we' do rely mostly on sight as a matter of historical and cultural contingency. If 'we' refers to an immutable human sensorium fixed over millennia both pre- and postmodernity, then this fails to take account of the history of the senses, their training, their impact on technologies and the impact of technologies on them, especially the ocular technologies of the railway, photography, cinema, car, television, computer monitor and smart phone (see Giblett 2008).

The sense of smell does later play an important role for Thurston (2004, 63) when he returns to Nova Scotia after a two-year absence in central Canada and crosses the Tantramar Marshes. He realizes that by smelling 'the odours of the marsh,' 'pungent, mephitic' he knew he was home, rather than by the seeing the marsh as the marsh watcher, or by hearing the sound of the willet as he did as a young boy at the Chebogue marsh and as an adult at the Old Marsh at Tidnish. *The New Shorter Oxford Dictionary* defines 'mephitic' as 'offensive to the smell; (of a vapour or exhalation) pestilential, noxious, poisonous.' In this regard it is close to 'miasma' defined in the same dictionary as '(an) infectious or noxious vapour, esp. from putrescent organic matter, which pollutes the atmosphere.' Yet the mephitic is offensive to the sense of smell and can be poisonous, whereas miasma are noxious without necessarily smelling bad and malaria is literally 'bad air' and a disease. The mephitic is certainly poisonous in Herman Melville's *Confidence Man* in which the disease (and figurative character) of 'Typhus,' 'taking his constitutional with Death [...] and three undertakers, in the morass, snuffs up the mephitic breeze with zest' (cited by Giblett 1996, 119). Typhus knows that the mephitic means more business—deadly business. By contrast with Typhus, Thurston snuffs up the mephitic marsh odours with zest in returning home to a healthy, life-giving, not an unhealthy, death-dealing place. These are more entries for an immersion in wetlands English.

Not only do sea and river water intermingle in the marsh for Thurston (1990, 90), but also for him 'the marsh appears to be the critical link between the three earthly domains—air, land and water—where they converge in the upper reaches of [the Bay] Fundy.' These three earthly domains are also three of the four traditional elements in European philosophy with the fourth being fire. The mixing of air, earth and water in the marsh is for Thurston a productive mixture and not the miasmatic and mephitic mixing of air and water in mal-aria or ague, nor the muddy mixing of earth and water in slime and the humour or affect of melancholy. Thurston (2004, 151) later likens the salt marshes to the water babies of Charles Kingsley's tale of that title 'inhabiting two realms, one in water and one in air.' Thurston goes on to reflect that 'humans also have a dual existence, one in solitude, which seems submarine, and one in society, which is more aerial.' The latter is certainly applicable to modern humans after the introduction of radio and with the advent of Wi-Fi who, as McKenzie Wark (1994, 121) puts it, have aerials, but not roots. Yet Thurston has roots in a particular place as we have seen. He is like a mangrove tree that has roots both aerial and watery.

Thurston (2004, 73) earlier describes 'a zone where water, air, sunlight, and nutrients mix' in 'the slurry at the surface of the [marsh] mud' that 'provides all the necessities for the production of life.' One of the necessities of modern life that this zone produced in the past is 'the petroleum sequestered in the Earth' as Thurston (2004, 74) puts. Petroleum is a relic and the heritage of the carboniferous era when swamps ruled the earth, when the world was wetland. Petroleum is swamp power that fuels the internal combustion engines of what could be called, following Lewis Mumford, carboniferous hypercapitalism. As coal and gas are also relics and heritage of the same era, it is also swamp power that fuelled the steam engines of carboniferous high capitalism and that still fuels the coal- and gas-fired electricity generators of today's carboniferous hypercapitalism.

9.2 Icarus, Falling of Birds

Thurston is acutely aware of the dangers and damage caused by the burning of fossil-fuels, including their waste in the production process. He and the photographer Thaddeus Holownia teamed up recently to memorialize and mourn the death of tens of thousands of migratory songbirds who were lured 'like moths to the flame' of a hundred-foot-high flare at the Canaport Liquefied Natural Gas plant in Saint John, New Brunswick in September 2013. With Canaport LNG on trial, the fallen birds were collected and held as evidence at the New Brunswick Museum in Saint John. After two long years, Holownia and Thurston finally gained access and told the grim story in word and image of the bird's fiery and infernal deaths (Thurston and Holownia 2022, 5). Fire is not a suitable element for birds.

Thurston relates in his poem that these birds were on their migratory flight path to wetlands in the north:

> to claim their summer homes
> in the wet woodlands
> (the darker, the wetter, the better),
>
> in the thickest tangles,
> in the swamplands,
> in the sproutlands, in the hemlock ravines,
>
> around the edges of swales and bogs [...]
> in mosquito-infested wastes [...]
> around the muddy rim of some woodland pond.
> (Thurston and Holownia 2022, 13–14)

'A myriad of birds' going about their usual business and flying to these their seasonal wetland homes in September 2013 did not reach them as they were drawn 'like moths to the flame' (Thurston and Holownia 2022, 25) of the flare and dropped out of the sky like Icarus falling:

> O blameless burnt little birds
> scattered on the ground, so that
> the flame keepers wept,
>
> Images seared into their brains
> on sleepless nights,
> on waking days. (Thurston and Holownia 2022, 25)

These images are now seared into the brain of the reader of Thurston's poem and the viewer of Holownia's photographs interspersed throughout *Icarus, Falling of Birds* who may well weep in mourning too at their harrowing vision in words and images of dead birds and for their deaths in an act of ornicide, the killing of birds. By using this title for the poem and their book, Thurston and Holownia are deliberately invoking the Greek myth of Icarus who flew on wings made of feathers held together with beeswax. His father, the master-craftsman Daedalus, who made the wings advised his son not to fly too close to the sun. The gung-ho hoon Icarus ignored his father's advice and the wax melted with devastating consequences.

Unlike mythical Icarus wearing his artificial wings of feathers who flew too close to the thermonuclear sun in an act of hubris and suicide, real birds with their actual wings of feathers flew too close to a petrochemical flare in an event of horror, ornicide and genocide. There was no master-craftsman present to give the birds advice and warn them about the dangers of flying too close to the flare. As a result, tens of thousands of birds belonging to twenty-six species in mainly the genera of wood warblers, one species of which is endangered, were incinerated and fell to the ground (Thurston and Holownia 2022, 5, 11). The LNG company concerned was convicted and fined in 2015 at a rate of less than $100 per bird killed, an obscenely low price to put on the death of an exquisite individual living being in ornicide. The fine in total was also an obscenely low price to put on genocide.

By using the title *Icarus, Falling of Birds*, Thurston and Holownia are also alluding to Pieter Bruegel the Elder's painting, *Landscape with the Fall of Icarus* and W. H. Auden's poem about it, '*Musee des Beaux Arts*,' in which he writes:

> About suffering they were never wrong,
> The old Masters [...]
> In Breughel's Icarus, for instance: how everything turns away
> Quite leisurely from the disaster; the ploughman may
> Have heard the splash, the forsaken cry,
> But for him it was not an important failure; the sun shone
> As it had to on the white legs disappearing into the green
> Water, and the expensive delicate ship that must have seen
> Something amazing, a boy falling out of the sky,
> Had somewhere to get to and sailed calmly on.

Canaport is a port as the name suggests so someone on a moored or passing ship with somewhere to be going to, or get to, may have seen (depending on the conditions) suffering birds with forsaken cries falling out of the sky. Humans living their usual lives and going about their normal business may have turned away 'quite leisurely from the disaster' of 'something amazing' of burning birds falling out of the sky, just as astonishing as a boy falling out of the sky. Holownia's photographs of burnt birds that would have been lying on the ground initially, and were later stored on another horizontal surface as legal evidence, are rotated and displayed in the vertical axis as if they are falling out of the sky in a disturbingly realist repetition of the horrific event itself.

Landscape with the Fall of Icarus is housed in the *Musee des Beaux Arts*, the Museum of Fine Arts, in Brussels. By contrast, *Icarus, Falling of Birds* is not a landscape, but a portrayal of gross acts of animal cruelty exhibited in the museum of coarse arts. It reminds me of the portrayal of the gross act of animal cruelty of St George killing the dragon exhibited in the statue outside the State Library of Victoria in Australia and the one in the old city of Stockholm in Sweden in a display of coarse arts in an open air museum (Figs. 5.1 and 5.2). *Icarus, Falling of Birds* depicts 'an important failure' and a gross betrayal of legal compliance and animal conservation. Thurston and Holownia (2022, 35) bear witness to, and speak the truth to power about, what they calls this 'tragic incident' (not accident) and 'ecological disaster.' A naked gas flare is a tragic incident waiting to happen, and it did on this occasion with disastrous consequences individually, generically and ecologically.

It was a disaster, not only for the birds, their species and genera, but also for the 'ecos,' the whole earthly home and earthly home economics. About the suffering of the birds, the new masters of Thurston and Holownia are not wrong, just as the older master of Colin Thiele was not wrong (as we saw in the introduction to the present volume). Thurston and Holownia did not turn away or look the other way, but looked with steadfast vision at the horror. *Icarus, Falling of Birds* is a contemporary environmental artwork in word and image exhibited in the museum of coarse arts and ecological disasters. This museum is the ecological equivalent of the various Holocaust, or Shoah, Museums about Nazi genocide located around the world with similar lessons to teach, memories to preserve, history to display and desire not to repeat.

One wonders what happens at the thousands of other LNG plants around the world, including the one in Western Australia located on the Burrup Peninsula and situated on the Australasian-East Asian flyway for migratory shorebirds and waders. Perhaps the flares of these coastal plants and those on the rigs out to sea are located close enough to the ocean for any burning birds to fall or be swept into the water without leaving much trace, just like Icarus. Out of sight, out of mind.

The Burrup Peninsula has other features, such as its famous rock art, but:

> the emissions from the flares appear to be eroding and flaking the petroglyphs [...] The Burrup petroglyphs are mute, beautiful, inscrutable. Some express an assured artistic sensibility. Some are heart-warmingly humorous, like the jovial, chubby 'first smiley face.' The animals are so faithfully executed that we can recognise the [thorny] devil or each species of turtle. (Bulbeck 2019)

The burning birds of Canaport no longer 'express an assured artistic sensibility' in song and flight. None are 'heart-warmingly humorous,' but are heart-achingly inscribed in the surface of Holownia's photographs where they are silent, dead, inscrutable, the dead surface of the photograph welded to the dead birds. No doubt the bird-lover can recognize the species in the photographically faithful execution of the faithless execution of the birds. The living art of birds singing and the ancient art of rock painting are suffering a similar fate in the flares of the fossil-fuel industry.

Thurston and Holownia speak in word and portray in image the truth of the disaster on behalf of dead birds and living wetlands who cannot speak for themselves. They write words and portray images for pro-paludal wetland culture. Wetland conservation is a matter not just of conserving wetlands as recreational places and for their 'ecosystem services,' but also of conserving them as habitats for the more-than-human species who frequent them and call them home for a longer or shorter time on their journey to and from them. These species include not only migratory songbirds, but also the migratory shorebirds and wader birds who traverse the various flyways of the world linking the feeding and breeding sites of marshes, mudflats and other wetlands (see Darby 2020). Wetlands are waterbird habitat. Wetland conservation is also a matter of halting and abolishing the fossil-fuel industry that directly destroy lives (such as in the Canaport case) and indirectly destroys lives, livelihoods and habitats through rising temperatures exacerbating fires and floods and raising sea levels as a result of greenhouse gas emissions.

9.3 Reciprocity

The three domains and elements of air, earth and water (but not fire) are the habitat for three animal kingdoms—birds, fish and land animals. Some land animals and some fish, cross from one domain to the other. Some animals, such as eels are, as Thurston (2000, 68; 2009, 54) puts it in 'Arcadia: The Marsh Suite,' 'hybrid/of fish and snake, half at home on land' and half at home in water, and so at home in the slime made of mixing land and water, solid and liquid. Thurston (1990, 8) describes how schools of estuarine fish 'swarm across the marsh at high tide like the shadows of clouds ghosting over the land.' Swarming creatures are spectral creatures of the black sun and black water as they belong to no one element, but cross from one to another. By crossing from one element to another, for a theology based on a hard and fast divide between the elements, they come under a biblical interdiction as an abomination (*Leviticus* 11:41). Swarming creatures are an abomination as they neither walk, swim, crawl nor fly, but slither on their bellies, and as they inhabit neither land, water nor air, but the interstitial (and intertidal in this case) zone between all three. Swarming creatures are spectral monsters of the black sun of melancholy and the black waters of wetlands as they belong to no one element but cross from one to another.

Yet the intertidal zone, and wetlands more generally, are for Thurston (2004, 5) 'the meeting place for Earth's two great biomes: the land and sea. It is this handshake, this biological reciprocity, that accounts for the richness of salt marshes, which are among the most productive habitats on the planet.' This handshake leads to a much more intimate intercourse and sexual exchange of fluids and nutrients in the wetland than Thurston is coyly inclined to suggest. He prefers to figure water as the lifeblood of the earth, such as when water separates from ice as 'a murky serum of brown, brackish waters' (Thurston 2004, 24). Serum, according to the dictionary, is the liquid that separates from a clot when blood coagulates and it is a therapeutic agent, especially in inoculation. Murky, brackish waters in Thurston's physiology of the body of the earth and of water as its lifeblood comprise the liquid discharged by melting ice that is a therapeutic agent for the land that inoculates it against diseases and promotes land health in Leopold's terms and mental health.

Reciprocity for Thurston (2004, 17) is 'our proper relation to the world.' He cites David Abram to the effect that this relationship is 'a "continuous dialogue" between the body and the things that surround it' and 'an active two-way conversation' rather than a passive one-way monologue. Biological reciprocity is a process that Thurston also explores under the rubric of mutual benefit. Thurston (1990, 88) suggests that 'where salt marsh and estuary are allowed to meet and intermingle, there is a mutually beneficial exchange of nutrients.' Similarly there could be more generally a mutually beneficial exchange between humans and placental wetlands of the Great Mother earth in the metabolic exchange of nutrients and wastes between them. Reciprocity has a long history in biological and ecological thinking, and makes a valuable contribution to it that Thurston briefly acknowledges without going into too much detail. These contributions include Eugene Warming's concept of symbiosis taken up and elaborated by Lynn Margulis in her concept of a 'symbiotic planet,' Jacob Moleschott's concept of *Stoffwechsel* (metabolic exchange) taken up and elaborated by Karl Marx, and Peter Kropotkin's concept of mutual aid taken up and elaborated by Murray Bookchin (see Giblett 2011, Chap. 12). All are useful for enriching and extending environmental sustainability biologically and politically away from anthropocentricity to biocentricity.

Reciprocity and metabolism do not necessarily imply an equitable and mutually beneficial relationship between both parties as one party can benefit to the detriment of the other. The parasite can kill the host. Every breath of air that we oxygen-breathing animals take affirms that humans are in symbiosis with the oxygen-producing plants of this planet. Modern industrial humans are producing more carbon dioxide than these plants can absorb and process into oxygen. In biological terms, a reciprocal and metabolic relationship operates on a continuum from parasitism through inquilinism (no harm or benefit to either party) to symbiosis. In Margaret Mahler's psychoanalytic terms, symbiosis is a psychological figure and process that concerns the mother–infant relationship on the same continuum. In psychoanalytic ecological terms, symbiosis, parasitism and inquilinism operate on a continuum in the human–Great Mother earth relationship (see Giblett 2011, Chap. 12; 2019, Chap. 9). In Mikhail Bakhtin's terms, language, literature and semiosis operate on a continuum between monologue and dialogue. Along similar lines, Thurston (2004, 17) develops a dialogic and bio- and psycho-symbiotic relationship to his 'home territory' of the bioregion of the eastern Tantramar Marshes in north-west Nova Scotia.

9.4 Tantramar Marshes

The Tidnish River in north-west Nova Scotia is located on the eastern side of the Chignecto Isthmus and drains its watershed north-east to Bay Verte, whereas the Tantramar River is located on the western side of the Isthmus and drains its watershed south-west to the Bay of Fundy. The Tantramar Marshes lie between. They are aptly described by Thurston (1990, 13; 2009, 90) as 'maritime prairie the sea laid down,' but were also, for better (or worse), 'known as "the world's largest hayfield"' (1990, 85) in their 'heyday' (89) (or should that be hayday?) at the turn of the twentieth century (Thurston 2004, 169). Thurston (1990, 80; 2004, 7 and 169; see also Summerby-Murray 1997, 164) translates the Micmac word Chignecto as 'great marsh district.' He relates that 'before the arrival of the Europeans, there were 20,000 hectares (or about 50,000 acres) of tidally nourished salt marsh at the head of the Cumberland Basin,' collectively known as the Tantramar Marshes. He goes on to relate that the marshlands of the Chignecto area comprising the Tantramar, the Missaguash, the LaPlanche and the Amherst marshes are collectively referred to as the Tantramar Marshes (Thurston 1990, 82).

The Chignecto Isthmus, in the words of Howard Trueman (1902, 12), 'separates the waters of the Bay of Fundy from those of Bay Verte and constitutes the neck of land which saves Nova Scotia from being an island,' a sad and sorry fate, perhaps worse than death, for those mainland and central Canadians who consider their islanders as quaint or backward or both. The neck of land that connects the head of Nova Scotia to the rest of the body of Canada is more precisely a wetland. This neck of wetland is the Tantramar Marshes. The marshes save the head of Nova Scotia from being disconnected from the torso of Canada as a backward island and the latter from being decapitated from the former as a headless monster. The marshes for Doucet (1999, 142) 'form the entire isthmus connecting Nova Scotia to New Brunswick.' The marshes are also the connective tissue between the head of Nova Scotia and the torso of the rest of Canada that supplies moisture and nutrients down its throat to the body of Canada. Rather than Nova Scotia being the head of Canada with the Tantramar Marshes as the neck, Nova Scotia for Thurston (2004, 171) 'barely hangs onto the continent by the Tantramar Marsh.' The marshes are the fingertips by which the body of Nova Scotia barely hangs on to continental Canada. The Tantramar Marshes also form what Thurston (1996, 1) calls a land-bridge between Nova Scotia and the rest of Canada, though it would be more precise to call them a wetland-bridge, more a causeway suspended in water than a suspension bridge hanging in air, a confederation causeway linking Nova Scotia to the rest of Canada, just as the 'Confederation Bridge' does for Prince Edward Island.

The marshlands of the Isthmus have played a role in the life and history of its peoples from First Nations through the Francophone Acadians to Anglophone colonists. Thurston (2004, 26) relates how:

> the oldest archaeological site in eastern North America was discovered shortly after World War II at a military base near the head of Minas Basin in the Bay of Fundy. Carbon dating tells us that Palaeo-Indians had camped there 10,600 years ago […] I like to think that these vanished peoples were the first marsh dwellers.

Or more precisely, they were the first marsh dwellers of the Fundy marshes. Aboriginal people were dwelling in the marshes and swamps of Australia at least 50,000 years before that.

The Acadians were the second marsh dwellers of the Fundy marshes. They dyked these marshes using *aboiteaux* (Trueman 1902, 16–17), as they did with Grand Pré (as we saw in the previous chapter). Thurston (1990, 68) indicates that the 'dyked land' of 'tidal meadows' was 'the basis of [...] Acadian culture.' Thurston (1990, 83) also notes later Acadia was 'a culture based on the technology of dyking.' More precisely, in Marxist ecological terms, the tidal marshes were the wetland foundation on which the economic base of dyked meadows and agricultural productivity employing the technology of dyking were built and, on which, in turn, the ideological superstructure of Acadian culture was erected. Without the foundation, base and technology, the superstructure would not have been possible. The economic base has an ecological foundation.

Thurston (1990, 86) relates and refers later to the fact that 'dyke land has been called "a reserve of energy in the form of fertility".' In the second section called 'Marsh' of 'Arcadia, The Marsh Suite,' Thurston (2000, 63; 2009, 49) invokes 'unkempt fertility: greenness, wetness.' The Acadians groomed, managed and drew upon that reserve of unkempt fertility deposited in the economic base for their culture that rested, in turn, on the earthly foundation of the fertile tidal marshes of the Bay of Fundy. An idea of just how extensive this economic and technological transformation was, and how massive the drawdown on the reserve undertaken largely by the Acadians and their successors was, can be gained by relaying Thurston's (1990, 82; 1996, 87) calculation that 'fully 90% of the original 357 square kilometres of salt-marsh shoreline in the upper Bay of Fundy was once dyked' and 'by the beginning of the twentieth century [...] converted to agricultural production' like a pagan savage converted from dark heathenism to enlightened Christianity. Dyking meant humans gained while other species lost as they have, as Thurston (2004, 123) puts it, 'suffered a loss of habitat due to dyking, draining, and infilling of salt marshes.' Humans drew on the energy of the marsh by taking it from the other inhabitants, rather than living bio- and psycho-symbiotic livelihoods in their bioregional home-habitat of the living earth.

Besides the problematic environmental history of the Fundy salt marshes, the Chignecto Isthmus has a shameful political history that Thurston barely touches upon. Perhaps as a nature writer he is not interested in, or concerned with, culture and cultural history, though the nature and culture of the Chignecto Isthmus are intimately intertwined as the *aboiteaux* attest. It seems difficult to acknowledge, Thurston, on the one hand the *aboiteaux* built and used by Acadians and not to mention on the other their Deportation in 1755 by the British from the same region, though Thurston manages to do so. The Chignecto Isthmus was where the final showdown between the French and English was played out in the late 1740s and early 1750s for control of the maritime region of Canada (Ross and Deveau 1992, 55–56) and where 'France made her last stand for the possession of Acadia' (Trueman 1902, 15). The Chignecto marshlands for Ross and Deveau (1992, 25, 55, 56 and 60) were:

> an extremely sensitive zone since the Treaty of Utrecht 1713 had transferred ownership of Acadie, or Nova Scotia, to the British [but] had left the western boundary of Nova Scotia ambiguous and undefined. France considered that it owned the territory north of the Missaguash River which cuts across the marshes of the Isthmus of Chignecto. The Acadians had been established in the settlement of Beaubassin on these marshlands for almost a century […] The contentious region of the Chignecto marshlands [was] where the two superpowers now faced each other, as if poised for war […] The first major conflict took place on the Isthmus of Chignecto at the beginning of June 1755.

The British were victorious and the Grand Dérangement or Great Upheaval, (or Great Deportation or Great Expulsion) began in 1755. The Chignecto marshlands were both a native quaking zone where land and water met for mutual benefit, and a feral quaking zone where two empires met for war, destruction and deportation.

The Chignecto Isthmus has a number of important wetlands that Thurston discusses in detail, including the Chignecto National Wildlife Area. This area is a Ramsar Convention Wetland of International Importance that includes John Lusby Salt Marsh and the Amherst Point Bird Sanctuary. Thurston (1990, 91) relates how the area for the latter:

> was acquired by the Canadian Wildlife Service (CWS) in 1968, and freshwater impoundments were developed there by Ducks Unlimited [Canada] in the early 1970s. It is unquestionably the most productive wetland in Nova Scotia. The diversity of habitat—wooded upland, wetland, gypsum sinkholes and abandoned farm fields—attracts a wide variety of bird species. In all, there have been 228 species sightings in the sanctuary, which ranks it second only [in the province] for the number of species sighted.

Not far from the Amherst Point Bird Sanctuary, and also part of the Chignecto National Wildlife Area, is the 600-hectare John Lusby Marsh that is, according to Thurston (1990, 82):

> the largest remaining salt marsh in the Bay of Fundy [...] As a measure of its importance, it was the first area to be acquired under the national land acquisition programme of the Canadian Wildlife Service (CWS), initiated to protect important wetlands in Canada [...] [I]n the 1960s it was a major staging area for Canada geese with as many as 15,000 honkers stopping here to feed for a month before proceeding to their breeding grounds [...] Now fewer than 1500 geese stop off in Fundy [...] [as] the Bay's salt marshes no longer present as good an opportunity for a meal as the winter wheat and stubble fields of Prince Edward Island.

Another wetland on the Chignecto Isthmus that Thurston discusses in detail is the Tintamarre Wildlife Area located on the Tantramar Marshes. Thurston (1990, 86) relates how:

> in the 1960s [...] the CWS created the 1,740 hectare Tintamarre Wildlife Area—a package of boglands, abandoned marshland and the Jolicure lakes—in the hope of increasing waterfowl production, which had been depleted by the intensive dyking and draining of the previous two decades. Ducks Unlimited [Canada] has been an active partner in developing the wetland habitat on the Tantramar, primarily by lending its technical expertise in building flood-control structures. Unlike *les aboiteaux*, the flood dams are designed to retain fresh water in an impoundment, thus creating better brooding habitat for waterfowl.

Yet, like the *aboiteaux* born out of European contact with the wetlands around the Bay of Fundy, these post-contact impoundments are not pre-contact estuarine wetlands with exchanges of salt and fresh water intermingled by the tide. Conservationists, according to Thurston (2004, 169), are 'returning some portions of the dyke lands to their natural state.' In some cases some are, as I saw below Fort Beausejour where the dyke had been breached and not repaired, but this is not the case in Tintamarre Wildlife Area that Thurston uses as an example of returning a wetland to its 'natural state,' presumably its pre-contact states, though he earlier related how the CWS created this area. Rather, wetland managers are here re-engineering dykelands as a hypermodern, or fourth, culture of nature as a conservation sanctuary within the larger Tantramar Marshes, most of which are still farmed. It is probably impossible to return these wetlands to the first culture of nature of First Nations as they were prior to the coming of the Europeans, partly because these people are no longer living on and working these wetlands and deriving their livelihoods from them, and partly because they have been so extensively modified by, and are still worked on by, the second agriculture of nature.

Tintamarre, the French name from which the Anglicized 'Tantramar' derives, means, as Thurston (2004, 39) points out, 'a great noise or din.' Yet rather than relating this to the noise of the river as most other writers do, Thurston relates it the great noise or din that 'the waterfowl wings made as the migrants rose over the salt marshes' as 'the Tantramar Marsh was a vital migratory feeding station for geese and other waterfowl along the Atlantic flyway.' The Acadians, as Thurston (2004, 39) goes on to argue, 'dyked those same marshes so many geese now overfly these once tidally nourished feeding grounds,' thus indicating the devastating impact the Acadians and their *aboiteaux* had on the marshland ecology.

Those impacts are now being reversed to some extent and the marshlands are re-establishing. Thurston (2004, 39) concludes that 'in recent times, dyke land is being allowed to return to the sea for the benefit of wildlife, society having decided the marsh has greater value in its natural state than as a hay and beef factory.' In recent times, dyked land is also being impounded for the benefit of wildlife (as I also saw below Fort Beausejour with Ducks Unlimited Canada's work, at Sackville Waterfowl Park and at the Tantramar Wetlands Centre), 'society' having decided that a pond has greater value in its cultural state than dyked marsh has as a hay and beef factory in a different state. Dykeland is also being allowed to return to the sea where it is impossible to maintain the dykes due to frequent incursions by the sea (as I also saw below Fort Beausejour). Dyked land is *not* being allowed to return to the sea where it is physically practicable, culturally advisable and financially viable to bolster it with rock (as I saw elsewhere below Fort Beausejour and at Westcock).

In the nineteenth century, the Chignecto Isthmus was subjected to two grandiose engineering schemes as part of the modern, or third, culture of nature. The ship railway (George Trueman 1899, map 92; Thurston 2004, 35–37) and the marsh canal (Howard Trueman 1902, map 3) were both designed to connect the Bay of Fundy with Northumberland Strait and the Gulf of St Lawrence across the Chignecto Isthmus (the Tantramar Marshes). Both the railway and canal are icons of early industrial capitalist modernity and both are transportation and communication technologies that conveyed goods and messages. The ship railway and the marsh canal are both represented as straight lines on father and son Trueman's maps. Both cut across the landscape of ridges and through the wetlandscape of floating bogs and black ash swamps. Both were designed to create a high way through or over the low lands, a straight and pure way through crooked and devious places, a clean and proper way through dirty and boggy places.

In the twentieth century, the Chignecto Isthmus was subjected to another grandiose hypermodern engineering scheme. When he considers the proposals for what he calls the 'grandiose scheme' of 'a perennial engineering' of building 'a tidal power dam' to 'exploit the high tides of the Bay of Fundy,' Thurston (1990, 91; 1996, 100) concludes that it 'would remove energy from the Bay's total budget by impairing the mutually beneficial link between sea and marsh.' The energy of the tides helps to generate the fertility of the marsh, as it would also generate electrical power. Damming that energy for electrical power would not only impair the mutually beneficial link between sea and marsh, but also enslave it. Moreover, like the earlier 'dyking, draining, backfilling or infilling and drying up' of marshes that Thurston (1996, 87) called 'this onslaught on wetlands,' damming the tide would be a further assault, the opening of a new front and the first thrust in another battle in the ongoing war against wetlands. War, as William Blake said, is 'energy enslaved.' The war against wetlands is the enslaving of their energy. This slave trade should be abolished and the slaves liberated. All ecosystems and all living beings require energy, and tidal marshes are one of the most energetic ecosystems on the living earth, hence their fertility.

9.5 The Future of Wetland Cultures

Hence the need to conserve wetlands now and into the future in order to sustain their energy-producing processes in and as habitats for humans and more-than-humans. Hence the need to acknowledge and respect the role of wetlands in the lives of more-than-humans and to halt and end the carboniferous hypercapitalist destruction in the fossil fuel industry of these beings and their homes. Hence the need to transition totally to renewable sources of energy that can help conserve these beings and recreate their homes. Hence the need to develop new and practical ways of conserving and appreciating all wetlands for humans and more-than-humans. Hence the need to continue to deconstruct and decolonize the mainstream western cultural denigration of wetlands as places of horror and its practices of dyking, draining, backfilling, embanking and infilling them to create farmland and suburbia. Hence the need to reconstruct wetland cultures so that all people (now and into the future) value, appreciate and conserve wetlands for humans and more-than-humans for what they provide for both and what they are themselves—places of exquisite aesthetic qualities and vital living processes. Hence the need to nurture nature writing that celebrates wetlands in all their facets and that bears witness to, and speaks truth to power about, the importance and value of wetlands to and for life on earth. Hence the need to nurture environmental artwork that portrays the exquisite nature of wetlands so all people on earth appreciate and care for them more. Hence, there is the need to read such nature writing, view such environmental artwork, value both and promote them among all people.

Hence the need to also redevelop traditional ways and develop contemporary ways of being and living bio- and psycho-symbiotic livelihoods in bioregional home-habitats of the living earth embracing wetlands. One way of doing so is to reconfigure the earth as body with wetlands as placenta and kidneys and water as the lifeblood of the body of the earth, all of which play a vital role in life-giving, life-sustaining and life-renewing processes for humans and more than humans. Humans must not treat wetlands as cesspools into which the sewers of rivers run, and not treat rivers as a digestive tract that process and excrete nutrients, but as part of a living body with other organs, such as the kidneys, womb and placenta of wetlands. Please help save or restore and protect wetlands. Long live pro-paludal wetland cultures!

References

Bulbeck, Chilla. 2019. Passion for the Burrup. https://greens.org.au/wa/magazine/passion-burrup. Accessed 31 May 2023.

Darby, Andrew. 2020. *Flight Lines: Across the Globe on a Journey with the Astonishing Ultramarathon Birds*. Crows Nest, NSW: Allen & Unwin.

Doucet, Clive. 1999. *Notes from Exile: On Being Acadian*. Toronto: McClelland & Stewart.

Giblett, Rod. 1996. *Postmodern Wetlands: Culture, History, Ecology*. Edinburgh: Edinburgh University Press.

———. 2008. *Sublime Communication Technologies*. Basingstoke: Palgrave Macmillan.

———. 2011. *People and Places of Nature and Culture*. Bristol: Intellect Books.

———. 2013. *Black Swan Lake: Life of a Wetland*. Bristol: Intellect Books.

———. 2019. *Psychoanalytic Ecology: The Talking Cure for Environmental Illness and Health*. London: Routledge.

Leopold, Aldo. 1949. *A Sand County Almanac and Sketches Here and There*. Oxford: Oxford University Press.

Ross, Sally, and Alphonse Deveau. 1992. *The Acadians of Nova Scotia: Past and Present*. Halifax: Nimbus.

Summerby-Murray, Robert. 1997. Interpreting Cultural Landscapes: A Historical Geography of Human Settlement on the Tantramar Marshes, New Brunswick. In *Themes and Issues of Canadian Geography III*, ed. C. Stadel, 157–174. Salzburg, Austria: Institute of Geography and Geoinformation, University of Salzburg.Salzburger Geographische Arbeiten, 34

Szabo-Jones, Lisa. 2011. Interview of Harry Thurston. *The Goose* 9: 80–93.

Thurston, Harry. 1990. *Tidal Life: A Natural History of the Bay of Fundy*. Camden East: Camden House.

———. 1996. *The Nature of Shorebirds: Nomads of the Wetlands*. Vancouver: Douglas & McIntyre.

———. 2000. *If Men Lived on Earth*. Wolfville: Gaspereau Press.

———. 2004. *A Place Between the Tides: A Naturalist's Reflections on the Salt Marsh*. Vancouver: Douglas & McIntyre.

———. 2009. *Animals of My Own Kind: New and Selected Poems*. Montreal: Signal Editions.

Thurston, Harry, and Thaddeus Holownia. 2022. *Icarus, Falling of Birds: A Poem by Harry Thurston, Photography by Thaddeus Holownia*. Jolicure: Anchorage Press.

Trueman, George A. 1899. The Marsh and Lake Region at the Head of Chignecto Bay. *Bulletin of the Natural History Society of New Brunswick* 4 (2): 92–104.

Trueman, Howard. 1902. *The Chignecto Isthmus and Its First Settlers*. Toronto: William Briggs.

Wark, McKenzie. 1994. Third Nature. *Cultural Studies* 8 (1): 115–132.

Index[1]

[1] Note: Page numbers followed by 'n' refer to notes.

R. Giblett, *Wetland Cultures*, Literatures, Cultures, and the Environment, https://doi.org/10.1007/978-3-031-57365-1

The manufacturer's authorised representative in the EU is Springer Nature Customer Service Centre GmbH, Europaplatz 3, 69115 Heidelberg, Germany. If you have any concerns regarding our products, please contact ProductSafety@springernature.com

Printed and bound by CPI Group (UK) Ltd, Croydon, CR0 4YY
01/10/2026
02225954-0003